MW01627952

HIGH VOLTAGE

To Eric
Here's to
another good
neighbor!

[illegible signature]

18/11/26

HIGH VOLTAGE

HYDROELECTRIC DEVELOPMENT AND POLITICAL POWER IN PERU

GONZALO ROMERO SOMMER

UNIVERSITY OF NEBRASKA PRESS ☰ LINCOLN

Chapter 4 was previously published as "The Changing Meaning of Cooperation: Rural Electrification in Cold War Peru, 1964–1976." *The Americas* 79, no. 4 (2022): 653–79. Reprinted with permission. Published by Cambridge University Press on behalf of Academy of American Franciscan History.

The University of Nebraska Press is part of a land-grant institution with campuses and programs on the past, present, and future homelands of the Pawnee, Ponca, Otoe-Missouria, Omaha, Dakota, Lakota, Kaw, Cheyenne, and Arapaho Peoples, as well as those of the relocated Ho-Chunk, Sac and Fox, and Iowa Peoples.

For customers in the EU with safety/GPSR concerns, contact:
gpsr@mare-nostrum.co.uk
Mare Nostrum Group BV
Mauritskade 21D
1091 GC Amsterdam
The Netherlands

Library of Congress Control Number: 2025027021

Set in Minion Pro by A. Shahan.

CONTENTS

List of Illustrations vii

List of Abbreviations ix

Introduction: An Electric Century 1

1. Fragmented Power 17
2. Vertical Limits 53
3. Electrification from Above 89
4. Electrification from Below 127
5. Electric Revolutions 163

Epilogue: From Light to Darkness 203

Acknowledgments 219

Notes 223

Bibliography 255

Index 277

ILLUSTRATIONS

PHOTOGRAPHS

1. Map of Peru 21
2. Members of the Sociedad Geográfica de Lima 23
3. Peru's centennial celebrations 41
4. Display of electric kitchens 43
5. Offices of Lima Light and Power Company 57
6. Callahuanca Power Plant 75
7. LL&P's bulletin *Kilowatito* 86
8. Vertical depiction of LL&P's hydroelectric chain 87
9. Peru's electric regions 104
10. The worker of the *sierra* 119
11. Huinco Power Plant 121
12. Peru's electric "take-off" 123
13. Evensen and fellow *sicainos* 147
14. Electricity arrives at the town of San Jerónimo, Mantaro Valley 152
15. Sicaya-Huarisca Power Plant 157
16. "Light comes to the peasants with the revolution" 157
17. Mantaro Plant zone of influence 169
18. Building the Mantaro dam 191
19. Electric nationalism 194
20. Mantaro Hydroelectric Plant 196

TABLES

1. Installed capacity, 1950–1965 (KW) 122
2. Installed capacity, 1968–1980 (MW) 199

ABBREVIATIONS

AEE	Asociación de Empresarios Eléctricos/Association of Electrical Businessmen
AEP	Asociación Electrotécnica Peruana/Peruvian Electrotechnical Association
AP	Acción Popular/Popular Action
APRA	Alianza Popular Revolucionaria Americana/American Popular Revolutionary Alliance
CAEM	Centro de Altos Estudios Militares
CEPAL	Comisión Económica para América Latina y el Caribe/United Nations Economic Commission for Latin America and the Caribbean
CPC	Cerro de Pasco Corporation
CORMAN	Corporación de Energía Eléctrica del Mantaro
DC	Democracia Cristiana/Christian Democracy
EDF	Électricité de France
EE. EE. AA./LL&P	Empresas Eléctricas Asociadas/Lima Light and Power Company
ELECTROLIMA	Empresa Regional de Servicio Público de Electricidad
ELECTROPERU	Empresa Electricidad del Perú
FEBO	Empresa Comercializadora de Energía Eléctrica de Muquiyauyo
FNDE	Fondo Nacional de Desarrollo Económico
GRFA	Gobierno Revolucionario de la Fuerza Armada/Revolutionary Government of the Armed Forces
GW	Gigawatts
HIDRANDINA	Energía Hidroeléctrica Andina
HP	Horsepower
IBRD	International Bank for Reconstruction and Development

INCOOP	Instituto Nacional de Cooperativas/National Institute for Cooperatives
INP	Instituto Nacional de Planificación/National Planning Institute
IPC	International Petroleum Company
ISI	Import Substitution Industrialization
JOP	Junta de Obras Públicas/Public Works Committee
KVA	Kilovolt-amps
KW	Kilowatts
KWH	Kilowatt hours
MDP	Movimiento Democrático Pradista
MG	Megawatts
MIR	Movimiento de Izquierda Revolucionara/Revolutionary Left Movement
NRECA	National Rural Electric Cooperative Association
ONDECOOP	Oficina Nacional de Desarrollo Cooperativo
ONDEPJOV	Oficina Nacional de Desarrollo de Pueblos Jóvenes
PC	American Peace Corps
REA	Rural Electrification Administration
SCIPA	Servicio Cooperativo Interamericano de Producción de Alimentos
SEIN	Sistema Eléctrico Interconectado Nacional
SEN	Servicios Eléctricos Nacionales
SGL	Sociedad Geográfica de Lima
SINAMOS	Sistema Nacional de Apoyo a la Movilización Social
SNA	Sociedad Nacional Agraria/National Agrarian Society
SNI	Sociedad Nacional de Industrias/National Society for Industry
SUDELECTRA	Compañía Sudamericana de Electricidad
TVA	Tennessee Valley Authority
UNO	Unión Nacional Odriista
USAID	United States Agency for International Development

HIGH VOLTAGE

Introduction

AN ELECTRIC CENTURY

July 28, 1980, was meant to be a celebratory moment in Peruvian history, as democracy returned to a country that had been under military rule for more than a decade. However, as the winner of that year's elections, Fernando Belaúnde Terry, delivered his inaugural speech, a high transmission tower located almost 5,000 meters above sea level in the central Andes was brought down by a powerful blast of dynamite.[1] Although of unforeseen significance at the time, the explosion was the first attack on the electric grid by the Shining Path, a Maoist terrorist group whose power and cruelty would grow throughout the 1980s, plunging Peru into a conflict that would claim nearly 70,000 lives. As the group intensified its actions, blackouts became commonplace and widespread, eventually affecting the capital city of Lima. In time the availability of light became a political question. If the lights were on, a night of relative peace and stability could be expected; if they went out, the fate of the state itself seemed to be in jeopardy. Politics had both literally and metaphorically become a matter of light and darkness.

The Shining Path had initially emerged in Peru's impoverished Andean provinces, a region in which Peru's leaders had hoped to enact change through the developmentalist potential of electricity for most of the twentieth century. The source of this power would be the region's rivers, which would be harnessed by clean, efficient, and modern hydroelectric plants. This process reached its conclusion with the inauguration of the Mantaro Hydroelectric Plant in 1973, a massive state infrastructure project located in the impoverished department of Huancavelica. But the electricity generated by the Mantaro system did not transform the Andes; rather, it ended up being consumed by the growing city of Lima, many kilometers away, on the Pacific coast. The Shining Path's attacks on Lima's supply of electricity, thus, were a stark reminder that the fate of Peru's capital coast was inextricably intertwined with the social conditions of the Andean interior, and that Peru's elites had ignored the underdevelopment of the Andes at their own peril.

Peru's experiences with electrification are a clear case of what scholars have described as "infrastructural politics."[2] As it is an often contested term,

I would like to contribute to this debate by defining infrastructural politics as a process in which high-level politics revolve around attempts to solve large-scale national problems via infrastructural efforts. In the Peruvian state, infrastructure had the aims of overcoming geographic and racial fragmentation, promoting economic development, and solving the political, social, and economic demands of its citizens.

Such infrastructural dynamics were by no means a new phenomenon in Peruvian history, as Lima-based elites had experimented with various infrastructures in the past. In the nineteenth century, elites had pinned their nation's developmentalist hopes to railroads, while in the early twentieth century, road building and aviation were championed.[3] While previous endeavors had been unsuccessful in achieving their goals, elites were certain that the technology of hydroelectricity was different. High in the mountains, hydroelectric plants would be built by Peru's Indigenous populations, and engineers and intellectuals believed that no human being in the world other than an Indigenous person could endure such grueling work. Geographic divisions and associated social and economic disparities would disappear, as all citizens were to enjoy the fruits of an industrial revolution thought by elites to be the natural outcome of its electrification endeavors. As in the past, it was thought that all the country's ills could be cured by the magic wand of infrastructure.

Peruvian elites' reliance on infrastructural politics proved to be a double-edged sword. On the one hand, political support for infrastructure yielded positive results. Peru constructed a large interconnected electric grid over some of the world's most challenging geography, and electricity as a public good was routinely consumed by its citizens. On the other, as the citizens of Lima were reminded by the actions of the Shining Path, this faith in the ability of electricity to solve Andean economic and social grievances not only failed but placed the very electric infrastructure built throughout the century at risk. The fault lay not in the infrastructure itself, as the sophisticated plants consistently produced and transmitted energy, but in the mistaken belief that infrastructure alone could be an alternative to real structural change. The fantastic promise of infrastructure, hence, was offset by its dangers and limits.

Peru's electrification process offers a vehicle to carry out an ethnography of the Peruvian state, analyzing the ideological and institutional dynamics of infrastructural politics.[4] Included in this analysis are the personal ambitions and political inclinations of presidents, ministers, representatives, and bureaucrats, who debated how much the state should intervene and what the goal of infrastructural development ought to be. The voices of regional elites are also at the forefront, as they supported hydroelectric infrastructures with the same enthusiasm as Lima-based governments, hoping to be treated as equals by decision-makers in the capital. It brings to the fore the role of engineers, who, while claiming scientific neutrality, became politically committed to hydroelectric development, holding key posts in both the public and private sectors, writing technical reports and feasibility studies that guided state actions. Not limited to the domestic arena, infrastructural development also had an impact on Peru's international relations, as hydroelectric development was characterized by moments of cooperation and dispute with great powers, international organizations, and transnational private actors.

To a lesser extent, this book explores the relationship between the state and citizenry through electrification. Most of the country's hydroelectric installations were built thousands of meters above sea level, resulting in isolated infrastructural endeavors that required limited negotiation with local populations, albeit with some exceptions. The limited interaction between the state and its peoples throughout the process of electrification may explain, perhaps, the growing gap in these projects between the state's technological prowess and its legitimacy with its own citizens, as well as the unfulfilled promises of Peru's electrification process.

THE VOLATILE POLITICS OF INFRASTRUCTURE

While this book seeks to offer a detailed account of the history of Peruvian hydroelectricity and its association with development, it has been informed by different theories that allow the story to be interpreted in different ways. The origins of this interdisciplinary focus are not hard to find, for, as of late, the study of infrastructure seems to have taken center stage. Seeking to understand the impact of physical infrastructures on people's everyday lives, anthropologists and others dove headfirst into the present experiences and future imaginaries that infrastructure holds. Present experiences are anything but harmonious, with physical infrastructures producing or reinforcing—intentionally or not—social, economic, and political segregation.[5] As for the future, they always promise progress and modernity, whatever those two concepts may mean at any given time.[6] In all cases, they are seen as tools of the state to regulate and control the lives of its populations. While it originally emerged from the field of urban studies, infrastructural politics, as defined in this book, seeks to complement these approaches by applying them at the national level.[7] By adjusting the scale, one can include the interaction among competing ideologies, heterogeneous populations, and complex geographies—as well as the transnational dynamics inherent in electrification systems in the Global South—to better understand these infrastructural promises.

Historical approaches can also benefit from another discipline that has made incursions in the field, although for separate reasons and targeting a different—more abstract—kind of infrastructure. Seeking to find historical explanations for the "weakness" of Latin American institutions, political scientists began studying how states have exerted their power over their territories in the past. Overall, these works have largely been influenced by the concept of the infrastructural power of the state, first developed by Michael Mann in the early 1980s, and subsequently rediscovered by Latin American scholars in the twenty-first century.[8]

Mann's concept of state power addresses the social capacity that states possess in exerting control and enforcing policy in their territories, a complex process in which physical infrastructures are only one factor among many.[9] However, Mann's work presents this infrastructural power as politically neutral, much as Max Weber's monopoly of violence was some hundred years ago. It is simply how the state operates. This book argues, however, that while physical infrastructure appealed to many because of its technocratic and apparently apolitical nature, it became politicized as multiple political movements, which sought to gain access to or influence the state, centered hydroelectricity in their own visions. This added another level of analysis to infrastructural politics. As power swung back and forth between governments of diverse ideological leanings, the deployment of this modern technology was aimed toward different political, economic, and social ends.

Indeed, statesmen, intellectuals, and engineers, as well as ordinary Peruvians, imbued hydroelectricity with their own hopes and dreams, prejudices and fears. Some paradoxically believed that electricity could hold the tide of modernity itself, strengthening social and regional hierarchies that had characterized everyday life for centuries. Others, hoping for change, saw in every plant a reformist and even a revolutionary means to dismantle those very same traditional structures. Electrification was a technical success as it physically connected large parts of the country, but it simultaneously highlighted critical divisions in Peru's political class, despite an almost universal belief in its transformative power.

A brief historical overview offers a clearer picture of the Peruvian brand of infrastructural politics. In the decades following the social and economic upheavals caused by the Great Depression, oligarchic and liberal administrations saw in electrification the promise of keeping in place social and regional hierarchies inherited from the colonial era. For Europeanized elites in Lima, electricity could ensure that they would retain their dominant position, as the potential industrial development of the central Andes was seen as an alternative to carrying out land reform. It could also reinforce racial and social hierarchies by slowing down Indigenous immigration to the "white" and *mestizo* capital, as it was believed that electricity would lead to the establishment of large industrial plants in the Andes that would halt the Andean exodus.

With the onset of the Cold War, reformist governments and even a "revolutionary" military regime reimagined electricity as a power for greater political decentralization and social inclusion long resisted by traditional elites. Together with land reform, electrification sought to improve the economic and social conditions of the Andean peasantry amid intensifying agrarian discontent. Given the push and pull of Peruvian politics, one may ask whether such political bifurcations weakened both oligarchic and reformist national aspirations. Complex infrastructures were erected, output increased each year, and electricity was consistently consumed. But revolutionary dreams of change failed, as Lima maintained its dominant political and economic position, industrialization outside the capital never materialized, and rural poverty persisted. But oligarchic dreams were also shattered, as the electricity generated by the Mantaro system illuminated the homes of millions of immigrants who left the countryside to live in Lima's surrounding shantytowns, changing forever the nature of the once colonial city. Geographic, racial, and political divisions remained unresolved, and, indeed, new challenges were created that future governments were unable to solve. The Peruvian state had expanded its infrastructure, but the unexpected outcome of its electrification endeavors raises the question of whether it had expanded its infrastructural power.

GEOGRAPHY, RACE, AND EXTINCT EMPIRES

Peru's volatile interaction with hydroelectricity goes far beyond its fascinating political narrative: it is not simply characterized by ideological disputes and political intrigues and, at times, personal ambition. This book also explores how the development of hydroelectric infrastructures was closely tied to how statesmen, intellectuals, and scientists understood Peru's geographic configuration and its linkages to efforts at state and nation building. Since independence, in the early nineteenth century, Peru's geography had become the subject of intellectual debates as an obstacle for achieving a politically and economically cohesive nation-state. The Andes, the country's dominant geographic feature, had long been portrayed by domestic elites and foreign travelers as an obstacle that fragmented Peru into three main geographic zones: the *costa*, *sierra*, and *selva*.

For much of the twentieth century, the costa, or coast, a thin strip of desert facing the Pacific Ocean, was home to Peru's Europeanized elites living in the capital city of Lima, who established dynamic agricultural and, to a lesser extent, financial and industrial activities in the region. The sierra, or highlands, refers to the Andean mountain chain. Its inhabitants were Indigenous peoples who worked the land under the direction of *gamonales*—regional landed elites and powerbrokers—as well as in the region's mines. As for the selva, or the Amazonian rainforest, it was depicted as an "empty space" whose imaginary and real wealth was to be exploited in the future. Because of its fragmented topography and Indigenous population, the Andes came to be depicted as "backward," a region intrinsically opposed to Peru's "modern" coast, and an obstacle to reaching the Promised Land of the Amazon. In the eyes of state and nation makers, such division presented a physical and social barrier that impeded the integration of the national territory.[10]

The above depiction will be familiar to those acquainted with Peruvian geography. Less known is how the development of hydroelectricity was essential in adding nuance to existing geographic debates, as hydraulic technologies depended on the rivers that rushed down the slopes of the Andes. The country's extreme geographic challenges transformed, under hydroelectrification, a traditional "obstacle to development" into what Albert Hirschman would have dubbed a developmental "possibilism," and, for a time, the Andes ceased to be a landscape upon which modernity had to be imposed and became one from which modernity could stem.[11] While the mountain chain no doubt remained a formidable barrier to national integration, including the construction of roads and railways, its dramatic altitudinal variations represented an "inexhaustible" supply of hydroelectric power. Because of their very natural constitution, the Andes impeded and, at the same time, held the key to Peru's development.

The topography of the Andes also meant that no large dams had to be constructed, as gravity was inherent to the mountain chain. No large reservoirs meant no real social engineering or large-scale displacement of peoples, therefore little local negotiation was required. This resulted in limited involvement by local actors in the process of infrastructural politics. Furthermore, the process of Peruvian electrification was based not on one grand emblematic project, but on a series of sophisticated plants placed at different altitudes, with the higher plants usually feeding the ones below. As such, the Andes, thus far relegated to the margins with their peculiar geographic configuration, offer a different avenue to exploring the dynamics of hydroelectric development, informing our understanding of other countries whose process of electrification was not based on one monumental undertaking.[12]

Andean imaginaries were tied not only to future possibilities but also to those of the past. As modern plants were embedded in the mountains alongside Inca roads and terraces, they became associated with the achievements of the former empire.[13] Hydroelectricity was a Western invention, but its Andean setting allowed for a renewed understanding of Inca ability to "harness" Peru's vertical geography.[14] Pre-Columbian civilizations had been able to create large empires amid the Andes because of their vertical conception of space and had been able to thrive in the mountains not in spite of geography, but because of it.[15] In this way, the developmentalism associated with hydroelectricity comes closer to notions of seeing Peru's landscapes in vertical terms, as opposed to the more horizontal approach found in modern road building.

This book argues that through hydroelectrification Peru was able, for the first time since independence, to find unity among its geography, Andean populations, and developmentalist aspirations. If ancient "Peruvians" had managed to build empires based on mastering geographic challenges, creating one of Karl Wittfogel's "hydraulic empires"—complex societies built around the need to manage water for irrigation—contemporary Peruvians could wield modern technologies and expertise to consolidate the nation based on an enduring tradition of overcoming geography as an obstacle to progress.[16] To loosely borrow a term from Alberto Flores Galindo, Peruvian elites were trying to "find their own Inca" through hydroelectricity.[17]

But pre-Columbian nostalgia also had political ramifications, for the "hydraulic empires" that Wittfogel spoke of represented not only an environmental reality but a political one as well. Inca greatness was due in part to their ability to manage their geography, but it was also possible because of their political structure, characterized by intellectuals as "despotic socialism." Such debates were not new and had been present since the beginning of the century, as some of Peru's most prominent thinkers invoked the Incas to support their own political vision for the country. Such debates only intensified in the polarized context of the Cold War, and politicians likewise had to grapple with the paradox that for Peru to achieve capitalist development through infrastructure it had to look for inspiration in the Inca's "socialist" past. As such, hydroelectricity also offers a novel way to explore how the global conflict was experienced in the local, particularly through contested interpretations of Peru's pre-Columbian legacy.[18]

THE STATE AND DEVELOPMENT

Hydroelectricity also allowed for the restructuring of the Peruvian state, as it dialogued with transnational ideas of development. Indeed, in many ways, the history of Peruvian electrification is the history of the Peruvian state's search for development. Electricity became synonymous with industry; industry equated with economic development. In time, the generation of electricity came to be seen as development in itself.

The origins of the electric industry in the early twentieth century were purely laissez-faire, but, given the sheer difficulty of building infrastructure in the Andes, state mediation was eventually required for large-scale projects. However, unlike other economies in Latin America, state intervention in the Peruvian economy to promote industrialization after the Great Depression was an inconsistent affair, and in the early years of the Cold War the state had abandoned any developmentalist ambitions.[19] Only in the mid-1950s, when considerable emphasis was placed on planning, did the state make a grand entrance. State intervention was facilitated by hydroelectricity, as national electrification plans were developed by the central government in partnership with foreign consulting firms. During the design and construction of the Mantaro system, capital and know-how flowed in from international agencies and various European consortiums. Peruvian bureaucracy followed suit by strengthening the Ministry of Fomento and Public Works through bureaucratic specialization and, eventually, creating a Ministry of Energy and Mines.[20] This bureaucratic buildup of the state reflected the impact of electricity, as the alliance between the state and transnational capital transformed government offices from social welfare–type services to a potential tool for national development.

Peru's experience with electrification coincided with modern ideas of development embraced by organizations like the World Bank. Infrastructure was king in this new developmentalist outlook, reaching its climax with Walt Rostow's "stages of economic growth" and Modernization Theory in general, which emphasized infrastructures as necessary for the economic "take-off" of "traditional" societies.[21] For these economic Cold War warriors, infrastructural development could also prevent the spread of communism by offering a blueprint to the Third World to compete with Marxist ideas of development. Peruvian statesmen might have been mistaken in thinking that infrastructure was the solution to all their problems, but all international trends reinforced this belief.

One might be even more generous and argue that, even by Latin American standards, Peru was characterized by a lack of political consensus regarding the appropriate avenue to achieving economic development, oscillating from liberal orthodox economic policies to reformist tendencies that called for greater state intervention.[22] The lack of consensus meant that the state's most consistent form of economic intervention was of an infrastructural nature and, in the face of resistance to development from Peruvian agrarian elites, the only avenue in which the state could exercise some degree of autonomy. In this sense, Peru differs substantially from the "classic" developmental states in east Asia, where alliances were established with industrial consortiums, rather than with infrastructural interests. As we shall see, this peculiar coupling will produce fewer tangible results than the clear process of industrialization that characterized the Asian "tiger economies."[23]

But Peruvians were not mere receptors of economic ideas, and this book seeks to engage with debates about the problematic concept of "development" in the context of Latin America's Cold War, which stresses the need to explore how the global conflict was experienced according to local realities.[24] Much of the current literature on Modernization Theory tracks its academic origins to the post–World War II American intellectual milieu. It also tends to be more telling of that milieu than of the reception of Modernization Theory abroad.[25] Furthermore, when addressing the theory's impact in the Third World, most scholarly inquiries on development have focused on the era of Cold War decolonization.[26] It is not, by any means, that Peru, or Latin America for that matter, did not go through many of the same dynamics as other countries in what we now call the Global South.[27] The public-private debate, the hopes of increasing living standards, the racialized nature of the electrification process, all these factors or variables were indeed present in the Peruvian experience. But unlike many of the newer postcolonial states in parts of Asia, the Middle East, or Africa, Peru had been imagining and implementing infrastructural programs since the nineteenth century, based on older, homegrown ideas of *fomento*, or national progress. These ideas had become institutionalized, and they informed Peru's interaction with Modernization and other theories of development (or underdevelopment). In this sense, while international ideas of development prioritized infrastructure, previous infrastructural experiences should have told Peruvian elites that the road toward development was not so simple.

TRANSNATIONAL FINANCIAL AND SCIENTIFIC FLOWS

Presenting an ethnography of the Peruvian state through the lens of infrastructure means exploring the intertwined relationship between the state and science. In this sense, Peruvian hydroelectric efforts were also highly transnational. Thomas Hughes, in his landmark work on electrical systems in Western society, stated that all electrical endeavors begin with the processes of invention and development of the new technology.[28] This means that in the United States and Europe the pioneers of electricity have rightfully received considerable attention, with Thomas Edison and Nikola Tesla taking center stage.[29] This has resulted in a gap that is beginning to be addressed by historians regarding the origins and development of electrical systems in the Global South.

In this sense, Hughes's second phase of development—that of technology transfer to other countries—is of greater relevance here. As electricity made an appearance in the Global South with its own geographical, political, and economic particularities, we do not find electric pioneers in the traditional sense, but "adapters" that applied a foreign technology to a society that had not invented it. Among these adapters we can find engineers such as the French-educated Peruvian Santiago Antúnez de Mayolo (1887–1967) and the Swiss Pablo Boner (1889–1972), both of whom applied a "Western knowledge" to a seemingly unique Peruvian landscape. In doing so, they came to resemble Stuart McCook's "creole scientists" of the nineteenth century.[30] "Creole" did not solely entail hybridity but referred to the rise of something distinctively Latin American—and, in this case, Peruvian. Both Antúnez de Mayolo and Boner also came to represent what Marcos Cueto has termed wielders of Andean knowledges, or *saberes andinos*, and may also be included among the ranks of those who engaged in hybrid scientific practices, leading to innovative research known as "scientific excellence from the periphery."[31] Including such figures in global histories of electricity allows one to explore how Peru, as well as other Latin American countries, did not passively absorb Western technology, but modified and/or adapted scientific practices in order to apply them to their particular contexts.[32]

Transnational scientific networks also went hand in hand with financial ones. Electrification was and remains a capital-intensive process, which requires the intervention of large financial actors to make it a reality.[33] In

Peru this transnational actor was the Lima Light and Power Company, a major player in this book. Thanks to its transnational linkages, the company was able to mobilize financial and scientific expertise. But such expertise was not politically neutral, and its representatives became embroiled in the game of infrastructural politics. Combining large amounts of Italian and Swiss capital, Lima Light and Power exerted considerable political power that went beyond sectorial interests, seeking to impose their own ideological notions upon Peruvian society, as the case of its fascist administrators during the company's early years will demonstrate. Other scientists, even the highly venerated scientific "wise man" Antúnez de Mayolo, lobbied and formed alliances with the state in their quest to promote ambitious hydroelectric schemes as well.

Such developments continued during the Cold War, as hydroelectricity became a variable around which ideological battles were fought and won. In the words of Andra Chastain and Timothy Lorek, "itineraries of expertise" were established, especially during the Alliance for Progress, a program that sought to promote economic development with the hopes of weakening communism in the aftermath of the Cuban Revolution. Experts aided Peru's Indigenous communities in the quest for rural electrification, only to find out that their expertise was quickly appropriated and modified by local actors. By bringing to the fore the role of such experts—both local and foreign—in Peru's electrification endeavors, the book once more seeks to understand how the global conflict was experienced by these figures on the ground, as they had to deal not only with regional and national politics but with local communities as well.[34]

A final word should be said about the inherently scientific nature of hydroelectrification, which is often presented as a totalizing project, as most hydroelectric plants in Peru bear the hallmarks of what James C. Scott termed "high modernist" projects. But Peruvian experiences with electrification problematize inherently negative perspectives about state planning and development, particularly those associated with high modernist ideology, which pushes states to make their territories "legible" through the simplification and uniformization of their geographies, environments, and populations via scientific and technical progress.[35] While Scott argues that such legibility destroys "local knowledges" that previously informed human interaction with their surroundings, the actions of Peru's Andean communities interacting with infrastructural projects suggest that this was not always the case.[36]

This book stresses that hydroelectric development, while not immune to the nature of high modernism, was not a wholly totalizing project, for statesmen, scientists, and populations alike combined the language of modern scientific progress with celebrations of the Inca past. While such celebration was at times superficial, in the case of rural electrification, particular forms of communal organization were linked to modern technological processes. This also goes against the grain regarding the link between science and development, for, despite Modernization Theory stating the need for Western scientific practices to do away with vestiges of tradition, tradition was compatible and even aided the arrival of electricity in the Andes. As such, universal ideas of science were not immune to local influences.[37]

THE SETTING: CENTER AND PERIPHERY

Peruvian electrification endeavors occurred on a national scale, but it was in Lima and the central Andes—which, for the purposes of this study, include the departments of Lima, Junín, and Huancavelica—where it was most intensely experienced and where it achieved its highest engineering form. The central region was full of hydraulic potential, being home to the Huarochirí and Marcapomacocha lake systems, which were diverted to the Santa Eulalia and Rimac Rivers, where transnational entrepreneurs built plants to light the homes of Lima's growing population and power small industries. Furthermore, part of the region was crossed by the Mantaro River, whose powerful waters allowed for the construction of the Mantaro Hydroelectric Plant, a feat yet to be repeated by the state and/or private interests. While references to other projects will be made—including the Santa Corporation in the north of the country—the central highland was the region where national dreams and ambitions were imposed upon with the greatest fervor.

The central Andes offers an ideal scenario to study the contested nature of infrastructural development for other reasons as well. The region was seen by statesmen as a corridor between the coast and the Amazon and was thus the main target of the developmentalist aspirations of the Peruvian state. As such, the region was already renowned for other impressive infrastructural achievements: first, the construction of the central railway in the nineteenth century, then the construction of the central highway in the early twentieth century. The study of hydroelectricity shows how new forms of infrastructure were imagined and how they were influenced—indeed, made possible—by previous infrastructural developments. Finally, the Mantaro Valley—the central highlands' most economically and politically dynamic region—has been widely studied because of its early exposure to capitalist development.[38] Hydroelectrification schemes, thus, became entangled with the existing economic, social, and political dynamics of the region.

More critically, many hydroelectric projects in the central Andes were built to supply electricity to Lima, which brings to the fore the uneven geographies of capitalist development by examining the tensions generated by the needs of urban cores as states attempt to develop their rural peripheries.[39] There is nothing new about this, as excellent studies have already shown how cities have developed at the expense of surrounding hinterlands. Yet many of these histories, including new histories of electrification, take place in the United States, relegating the state to a secondary role in the context of unrestricted capitalism. This reality does not apply in Latin America, where the state has at times been considered an agent of capitalism.[40] Furthermore, the setting allows the inclusion of rural populations that were impacted by electrification, especially in the Mantaro Valley, offering a counterpoint to most works dealing with electrification, which have disproportionally focused on urban settings.[41] Given that the majority of Peru's population lived in rural areas until the second half of the twentieth century, the study of rural electrification is of great relevance to understanding state attempts to bring development to the countryside, where there was always latent discontent.

Finally, this book does not solely focus on the uneven development between city and countryside but also explores the political consequences for the state as it attempted to promote capitalism in peripheral areas. This means bringing to the forefront the voices of regional elites, as influential figures in Junín, centered around the city of Huancayo, challenged Lima's claim to dominance, shining a light on how peripheral cities seek to challenge the position of a dominant city in a highly centralized state. As we shall see, the tension between center and periphery often expressed itself more forcefully through infrastructural disputes, as it appears that not only Lima-based elites, but regional ones as well, were taken by the promise of electricity.

ORGANIZATION

To write an infrastructural history presents some peculiar organizational challenges. Infrastructural projects outlive the governments that design, construct, and inaugurate them. Indeed, one of this book's arguments is that the meanings and imaginaries associated with infrastructural projects changed wildly over different political regimes during their construction. This suggests two organizing strategies. First, one could write a history following the standard political chronology of twentieth-century Peru. However, this strategy leaves infrastructural projects unfinished, and the reader would have to wait to find out about their culmination in later chapters. Instead, each chapter—except for the first—focuses on a particular infrastructural undertaking, offering a "biography" of a project by tracing it from start to finish, tying it to the larger political, social, and economic dynamics of the period. These biographies have been, to the extent possible, arranged chronologically, hopefully avoiding unnecessary repetition.

Chapter 1 explores the impact of hydroelectricity in the evolution of geographic narratives in the late nineteenth and early twentieth centuries, as intellectuals imagined the economic and social potential that the new technology offered in what was considered an "exceptional" geographic setting. Such narratives were quickly politicized with the onset of the Great Depression, as a new generation of provincial intellectuals began to use infrastructural arguments to criticize Lima's dominance over the rest of the country. Clamoring for greater political and economic decentralization, these young intellectuals argued that most of the nation's hydraulic resources were scattered throughout the Andes, strengthening their claims that Peru's diverse geography was not compatible with a highly centralized government.

Chapter 2 traces the rise of Lima Light and Power, Peru's largest electric company. From the early years of the twentieth century to the end of the Second World War, Lima Light and Power used Italian and Swiss capital to erect a sophisticated hydroelectric system, which took advantage of the violent changes in altitude in the Andes near Lima. However, during the interwar period, the company became involved in Peru's volatile political scene, experiencing a troubled relationship with its workers, as Italian administrators attacked Marxist-inspired labor unions and openly expressed fascist sympathies. Subsequently, Lima Light and Power's domestic troubles acquired international dimensions, as Peru designed a foreign policy in support of the Allied powers, resulting in a greater politization of hydroelectricity that symbolized the transnational struggle between the dominant ideologies of the time.

Chapter 3 is centered around the construction of the Huinco Power Plant, carried out by Lima Light and Power at the peak of its influence. The construction of the plant symbolized a greater interest on the part of the state in electrification endeavors, as Huinco required not only a fantastic engineering effort, which pushed the electric frontier high into the Andes, but also legal mechanisms and an ambitious national electrification plan, all of which allowed the World Bank to finance its construction. Construction of the plant was made possible by using the latest engineering innovations and because of the "man of the sierra," the only one who could work at high altitudes. However, the Huinco case shows that state support for electric infrastructure was not accompanied by an effective industrialization policy, leading the Peruvian state to become closely associated with an electric company as opposed to a purely industrial entity.

The politics of rural electrification are explored in chapter 4. Examining the electrification of the Mantaro Valley through the creation of the Eléctrica Comunal del Centro Ltda. No. 127 this chapter argues that rural electric cooperatives allowed Peruvians to channel traditional communal practices through an institution that was seen as modern, capitalist, and Western. Because it was aided by American cooperatives during the Alliance for Progress years, Peru's rural electrification endeavors rescued ideas and values espoused by the New Deal almost three decades after the program's arrival in the United States. As with other electrification efforts, electric cooperative development also became a heated political battleground. First presented as a viable way of promoting communal capitalism, it was eventually described as an obsolete institution by the Revolutionary Government of the Armed Forces, which took power in 1968. This government adopted its own cooperative model, free from any capitalist "vices," as its sought to implement its own "revolution from above," ultimately resulting in the Cooperativa Comunal del Centro's demise.

The last chapter traces the construction of the Mantaro Hydroelectric Plant, which spanned over three decades from its inception to its inauguration. First designed in 1940, the Mantaro plant was negotiated by a liberal administration, constructed by a reformist government, and inaugurated by a "revolutionary" regime, the political significance of the project undergoing radical changes. First envisioned as necessary infrastructure to spur industrial development and create the necessary conditions for Peru's "take-off," by the time of the plant's inauguration it was depicted as a radical tool to fight underdevelopment. The plant's construction also highlighted the challenges of Peruvian centralism, as regional elites in the central Andes bitterly fought for the project to go ahead. While the project divided Peru's political class, it also connected the country by creating Peru's first interconnected grid, absorbing the infrastructural achievements of the previous half century of often contradictory national development.

1

Fragmented Power

In his 1919 book *Elementos de geografía científica del Perú*, Óscar Miró Quesada lamented that Peru was a fragmented nation. "Because of the Andean Mountain chain, Peru is divided into three parts, which are akin to three different countries, because of their appearance, their climate, their animals, their plants and their inhabitants." Basing his work on the tripartite division of the country that had gradually emerged after independence, Miró Quesada was speaking of Peru's three "natural" regions: costa (coast), sierra (highland), and selva (rainforest).[1] While the costa and selva were generally plains with mild climates, Miró Quesada considered the sierra a geographic challenge, as the terrain of the Andes, while rich in resources, was "uneven and mountainous," as well as "rough and exceedingly cold."[2] Together with the fact that the Andes were largely inhabited by Peru's Indigenous populations, the region was depicted as a geographic and social barrier toward political unity and economic growth.

This opening chapter explores the linkages between geographic discourses and hydroelectric infrastructures during the first half of the twentieth century. Despite Miró Quesada's pessimistic depiction of the Andes—a view held by many of Peru's coastal elites—the arrival of hydroelectricity allowed geographers and other intellectuals to attempt to decipher the mountain chain's opportunities and challenges, ultimately considering that the country had an "exceptional" geographic setting for the hydraulic generation of power. As geographers debated back and forth, the paradox of Peruvian geography became increasingly clear, as this hydroelectric potential depended on the very geographic configuration that made Peru a fragmented nation in the eyes of Lima elites.

This renewed understanding of the Andes by domestic intellectuals was reinforced by transnational developments. Foreign travelers and scientists who passed through or settled in Peru early in the twentieth century celebrated Andean hydraulic potential. Literature on foreign travelers has tended to focus on nineteenth-century naturalists and their role in the geographical and social creation of the Andes, which resulted in its depiction as an obstacle to national unity and development.[3] However, twentieth-century scientists offered an alternative view, for "white coal" could transform the Andes into a region that would uniformly offer opportunities for industrial development. As their works were disseminated by diplomatic, scientific, and commercial publications, hydroelectric imaginaries ensued. They reflected what anthropologists have described as the "promise of infrastructure," as they emerged before most of the country's hydroelectric facilities were built.[4] While the reality of hydroelectricity did not always live up to its promise, it was nevertheless crucial in reimagining the Andean mountain chain as a source not of backwardness, but of potential modernity.

While the benefits of hydroelectricity were still largely a promise, they also symbolized the arrival of infrastructural politics as imaginaries were quickly politicized by Lima's elites. For Peru's upper classes the promise of hydroelectricity went far beyond dreams of economic modernization, as it could also solve myriad social questions. Electrification could "solve" the nation's racial problems, as Peru's Indigenous population—viewed negatively by Europeanized elites—could become "useful" via the merits of a technical education, or it could stop "undesirable" Chinese migration by multiplying the productivity of local workers. The surveying of hydroelectric potential also reflected glimpses of nationalism in an era in which knowledge from abroad was celebrated, as some of the elite believed that Peruvian engineers should be given opportunities equal to those of foreign ones. Finally, electricity could also act as a conservative force by weakening the nascent feminism of the time, as women could rediscover the domestic sphere with new appliances.[5] Hydroelectric possibilities, hence, were as much social and political as economic.

This politicization of hydroelectricity was by no means monopolized by the capital's elites. For regional intellectuals, hydroelectricity allowed them to imagine a political and administrative reconfiguration of the Peruvian state. Based in the capital city of Lima, the state did not have the institutional tools or reach to exploit the country's hydroelectric resources, as it lacked a proper division of labor—specialized institutions or agencies—a key aspect of the infrastructural power of the state.[6] Because of the inadequacies of this political arrangement, younger regional intellectuals could use infrastructure to put forward arguments for the country's decentralization and state expansion, which contrasted sharply with the aspirations of Lima elites, who thought that infrastructural development ought to be directed from the capital. Hydroelectricity, hence, would not only power Peru's potential industrialization but also energize old political battles between Lima and the rest of the nation.

A MOST HARMONIOUS GEOGRAPHY

In 1883 Peru was a defeated country. Having lost the War of the Pacific (1879–1883), its infrastructure lay in ruins, its capital occupied by Chilean troops, and a parade of *caudillos* vied for power. Only a few decades earlier, the country had experienced an economic boom thanks to the discovery of guano deposits on a handful of islands off the southern coast. Yet the guano years, as Jorge Basadre noted, were of a "fictitious prosperity."[7] As the state became dependent on this commodity, massive debts were accrued, as foreign lenders paid large monetary advances for the fertilizer. Soon, deposits were nearly depleted. Guano's successor, the nitrates located in the southern provinces, would end up in Chilean hands after the war. It would take a decade for Peru to begin recovering from the consequences of the debacle.

Peruvian elites experienced a resurgence only when the country entered the highly unequal but economically dynamic period known as the Aristocratic Republic (1895–1919).[8] While still dependent on exporting raw materials to Europe and the United States, Peru's economy was more diverse than in the nineteenth century, including both agricultural and mineral products. Amid this newfound optimism, intellectuals struggled to come to terms with the defeat against Chile and understand why Peru lost the war. Initially, blame was placed on the Civilista Party, who had defunded the armed forces in its quest to put an end to the political dominance of military caudillos. Others lamented the folly of an alliance with Bolivia, which had dragged a reluctant Peru into the war. In time, however, the consensus emerged that, unlike Chile, Peru had been unable to build a cohesive national identity because of its Indigenous population, giving birth to what would be called the "Indian question."

In time, the inability to construct a viable body politic was explained by the two separate but interconnected factors of geography and race. As Benjamin Orlove has argued, with the foundation of the Republic, colonial notions depicting a diverse territory in which mountains and lowlands coexisted as part of a divine "cosmos" were gradually replaced with the three "natural" regions: the coast, the highlands, and the rainforest. (See figure 1.) This shift from a "colonial" to "republican" geography, which replaced disparate mountains with a monolithic highland, was a complex process in which both foreign scientists and the Peruvian state played their part.[9] The work of the German naturalist Alexander von Humboldt, who visited Peru in 1802, was key in this new geography, as he arrived with new tools like the barometer to measure elevation. The work of Italian geographer Antonio Raimondi was also of great importance, and his surveys were essential in the elaboration of Peru's first atlas, published by Mariano Felipe Paz Soldán in 1865 by order of the Peruvian state.[10] As the modern geographical conception of the Andes came into being, republican elites soon began depicting the complex and fragmented topography of the mountain chain as a barrier or obstacle that hampered Peru's quest for national integration.

This geographical reordering also had consequences as to how the state viewed its population. In colonial times who was or was not an "Indian" was determined by a complex legal and cultural system. After independence such distinctions were abolished, but liberal citizenship failed to take root with the emergence of a republican government. With the tripartite division of the territory, the state began conceptualizing its Indigenous population in geographical terms. The Andes became synonymous with the Indian and vice versa.[11] If the Andes were an obstacle to national integration, so were Peru's Indigenous populations, and intellectuals began to speak of the "Indian question" in geographical terms. The arrival of republican geography had allowed the Peruvian state to order and categorize not only its territory but also its peoples.

Despite the push for geographical homogenization, by the end of the nineteenth century Peru's tripartite division had given way to more dynamic interpretations. One need look no further than the bulletins published by the Sociedad Geográfica de Lima (SGL). Founded in 1888 the SGL was an institution of its time, reflecting international trends that were giving rise to a "geography of the state," in which geography and state power went hand in hand.[12] Its self-declared purpose was to study the country's geological constitution, map its agricultural and mineral riches, and further the state's knowledge of its territories. It also sought to portray Peru in a positive light to attract European migrants. Luis Carranza, one of the SGL's first directors, stated that its mission was to "amend the grave misconceptions that superficial observers have propagated abroad . . . presenting our climate as inappropriate for the acclimatization of the European race."[13] By portraying Peru as a suitable destination for European migrants, the SGL sought to solve Peru's "Indian question" through geographic knowledge.

The articles published by the members of the SGL revolved around two main debates. The first centered on which of Peru's regions showed more potential for future development, the Andes, with its mineral and hydraulic riches, or the selva, with its untapped and almost mysterious wealth. The

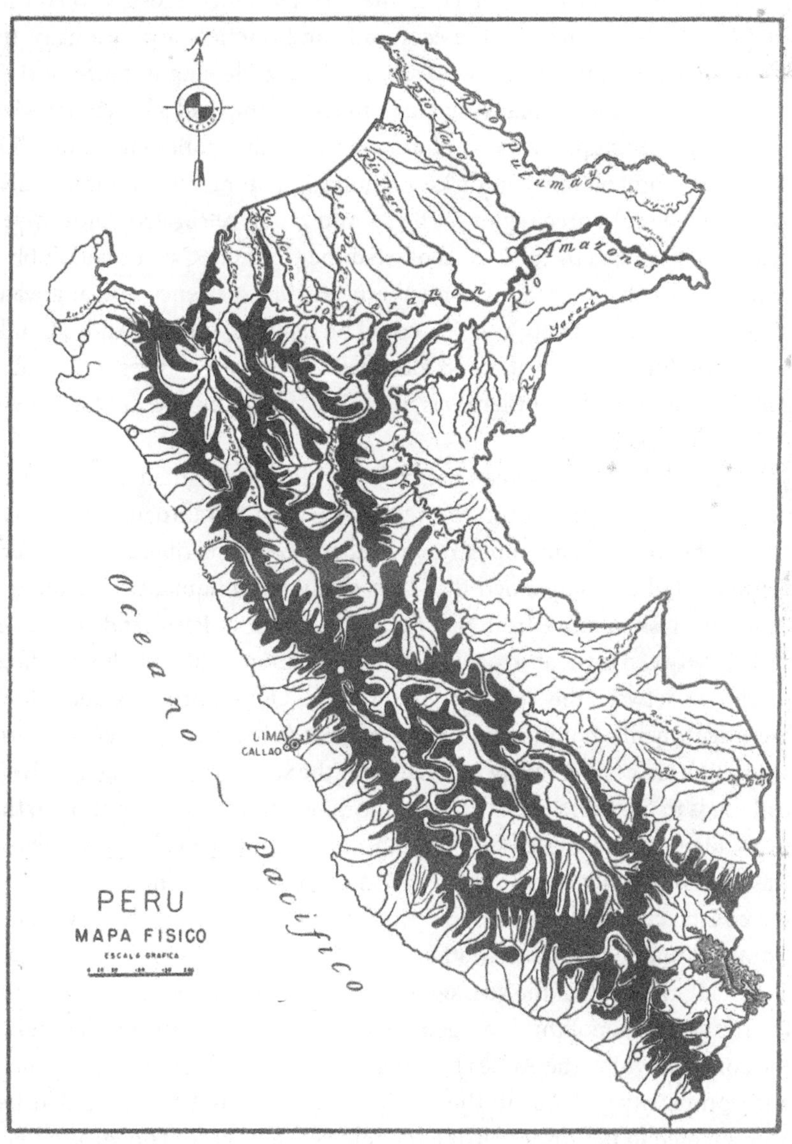

FIG. 1. Map of Peru, the Andes as a "barrier." José Pareja Paz Soldán, *Geografía del Perú: Manual* (Lima: Lib. Internacional del Perú, 1955). Pontifica Universidad Católica del Perú Libraries.

second debate was closely linked to the first and pitted those who considered the Andes a curse on the economic and political aspirations of the Peruvian state against those who saw them as a blessing because of their complex topography. Those situated in the first camp had a horizontal view of Peruvian geography, a view reinforced by the difficulties encountered in the construction of transport or "communication" infrastructures that sought to connect Peru's three regions. The second group celebrated Andean verticality, epitomized by the arrival of what the Count of Cavour had dubbed "white coal." This reflected the northern Italian experience of using water to generate electric energy, which could be replicated in the Andes thanks to the combination of water and altitude.[14] These two debates occurred in parallel and showed that there was no consensus on the possibilities offered by Peruvian geography.

Geographers and statesmen had for some time highlighted the challenges that Peru's geography posed to the construction of railroads and, subsequently, highways. In his popular 1912 textbook *Geografía comentada del Perú*, Germán Stiglich complained that "it is always an inhumane, difficult, and expensive task to move from one point to another in Peru" and concluded that "if between town and town all communication is difficult, how difficult must it be between the sierra and the coast!"[15] Óscar Miró Quesada echoed these sentiments, pointing out the lack of roads between the sierra and the selva. "We are commercially isolated from the selva and its navigable rivers; and roads in the sierra are so bad that the difficulties of transportation in that rich and populated area of Peru are almost insurmountable."[16] Such isolation was an obstacle not only to the creation of a dynamic economy but also to the construction of a nation, for peoples remained geographically distant from and unknown to one another.

Not all members of the SGL agreed with this analysis. Rather than seeing the sierra as a monolithic bloc, geographers pointed to Peru's borderlands—the zones in which the Andes began and ended—and found heterogeneity and complexity.[17] It was in these borderlands where the waters that had accumulated in the Andes during the rainy seasons flowed down and made life possible on the coast, irrigating large tracts of land and increasing its agricultural exports. But thinking of water as "white coal" also led to dreams of industrialization. Inspired by European developments in the field of hydroelectricity, former congressman and intellectual Ricardo García Rosell stated in 1893 that Peru should follow the example of countries such as Italy, France, Belgium, and the Netherlands, all of whom obtained from their rivers "the

FIG. 2. Members of the Sociedad Geográfica de Lima, with Óscar Miró Quesada on the bottom left, and Emilio Romero in the center back row. Sociedad Geográfica de Lima, *Cincuentenario de la Sociedad Geográfica de Lima* (Lima: Librería e Imprenta Gil, 1937). Biblioteca Nacional del Perú.

indispensable motor force to power their workers' activities."[18] García Rosell pointed out that Peru, with its pre-Columbian hydraulic tradition, could achieve such results as well, but modern Peruvians, much like the Incas had done in the past, would need to learn to take advantage of the hydraulic opportunities offered by the Andes.[19]

The importance of the Andes, hence, lay in its rivers. Peru was blessed not only because of the myriads of rivers that traversed the national territory but because they were "Andean," which imbued their flow with a powerful and violent force. A new breed of liberal economists, led by figures such as Alejandro Garland, would state in the bulletin of the SGL that "the amount of hydraulic force which can be provided by these rivers . . . which we have termed 'Andean,' is immense and can be advantageously applied in various ways, such as the electrification of railroads and many industries."[20] New technological developments could even lead to the revival of past glories, and Garland touched a nostalgic note when he considered that electricity could bring Peru back to the days of nitrate through the manufacturing of artificial saltpeter.

The hydraulic potential of the Andes was portrayed as a "trade-off," as the generation of hydroelectricity offset many of the challenges that the mountain chain presented to the building of communications lines. It was Eulogio Delgado, future president of the SGL, who most eloquently expressed this idea: "Our uneven territory is a grave inconvenience for easy and cheap roads, but they provide us with an inexhaustible supply of hydraulic force with which to produce electricity at low cost."[21] The unevenness of the territory was a blessing and a curse. It was a curse for the building of roads, which was portrayed as a herculean task. The same unevenness was a blessing, however, as hydraulic power was so abundant, argued Delgado, that it could supplant the use of oil, which, because of the very same lack of roads and railways, was highly expensive.[22]

For others, the challenges of Andean topography were not worth the cost, and not everyone was convinced of the virtues of "white coal." Engineer and former politician Enrique Coronel Zegarra argued that Peru had been dealt a rotten geographical hand, for the economic future of Peru was to be found not in the Andes but in the selva—an eternal "El Dorado"—including its *pongos*—from the Quechua meaning "door"—the narrow gorges that Andean rivers passed through on their way to becoming long and wide Amazonian waterways.[23] The potential "colonization" of the Amazon—a dream that was to captivate statesmen for much of the twentieth century—was denied to the fragile Peruvian state by the Andes, for despite "the minerals that exist within it, the hydraulic power that it currently holds and the variety of climates that it offers, (it) has constituted . . . an almost impassable obstacle for commerce between the Pacific and the (Amazon) river basin."[24] Even those who recognized the hydraulic potential of the Andes felt that it should be used to overcome the mountain chain and colonize the distant selva through European migration. Diplomat Federico Alfonso Pezet saw in hydroelectricity a way to decrease the operating cost of the central railway, essential for transporting European colonists to the other side of the Andes, as the region's "infinity" of streams "will easily generate enough force for electric locomotion at a low price."[25] Andean hydroelectricity could facilitate the construction of other forms of infrastructure, but its ultimate purpose was to link the coast and the Amazon basin and bypass the mountain chain altogether.

The debates surrounding the opportunities offered by Andean topography emanating from the SGL remained largely abstract in nature and, aside from occasional references to the infrastructural prowess of the Inca Empire, did not address the role to be played by Peru's Indigenous peoples. But geography also permeated the work of intellectuals outside the SGL, particularly those of the Generación del 900, a group of Lima-based oligarchic intellectuals born during or after the War of the Pacific and influenced by positivist thinking. Heirs to the broad-thinking *pensandores* of the nineteenth century, members of this group considered the future of Peru to be inevitably tied to the Andes and its peoples, and their views on Peruvian geography mixed elements of both the colonial and republican.[26]

Such was the case of *El Perú contemporáneo*, the landmark work by Francisco García Calderón. Published in 1907 the book inevitably deals with geography in its analysis of "Peruvian reality." For García Calderón, there were indeed three "well defined" regions, but he argued that their existence meant that Peru had a geographical diversity that simultaneously presented a challenge to achieving national unity and offered unbound economic potential through the appropriate use of scientific knowledge.[27] The very complexity of Peruvian geography set it apart from its counterparts in the South American neighborhood. It was the lack of the well-defined geographic configuration possessed by other countries that made Peru exceptional. "(Peru) does not have the vast plains of Argentina, or the lustful tropical monotony of Brazil, nor Chile's narrow extension." A "simpler" geography might aid in the process of national unity, but it made such nations less dynamic in the long run. The "diversities and oppositions" existent in Peruvian geography presented an unlimited productive potential, leading García Calderón to conclude that, in all South America, Peru was "the most harmonious country in its geographic variety."[28]

Influenced by existing social Darwinist notions of the time, García Calderón's celebration of Peruvian geography was not extended to its present inhabitants, but to those of the past.[29] He believed that Peru's Indigenous masses were inferior to its Europeanized elites, an inferiority that was the result of the Andean environment. García Calderón considered the Indigenous population, products of the mountain chain, to be slow and stubborn, much like the Andes themselves. Likewise, he believed that their "creative drive" was overwhelmed by their "imitative faculties." The lack of individual initiative did not prove to be an obstacle in pre-Columbian times, for under the all-powerful eye of the Inca they collectively built a great empire characterized by sophisticated infrastructural achievements. The greatness of the Inca Empire was caused by many factors, but perhaps none more so than geography itself, as "the climate, the action of the environment, favored the collective spirit and the physiognomy of the inferior classes (of the empire)."[30] Now, under the paternalistic direction of a republican government, Indigenous peoples could again play a part in Peru's resurgence. For this, it was necessary to establish an industrial education—the humanities were reserved for the "creative" oligarchy—as Peru's Indigenous population would benefit greatly from practical courses on electricity, mechanics, design, and agriculture, required to duly exploit the diverse economic potentialities that the country offered.[31]

Such tools were in accordance with the possibilities for development presented in early-twentieth-century Peru, as its "harmonious diversity" allowed for great economic versatility. García Calderón makes special mention of the electric industry, because of both its economic potential and its political implications. Lima Light and Power, a newly created electric trust, which will be explored in the next chapter, showed the power of combining liberal individualism and association, for the "union of interests produced by this collaboration, is the basis of peace and prosperity."[32] The last word on the electric industry, however, was not to be had by the will of industrious citizens, but by the nature of Peruvian geography itself. "We hope that the waterfalls, so abundant in our country, determine their great progress."[33]

El Perú contemporáneo, hence, stressed that the main obstacles to Peru's progress were determined not by geography but by the racial composition of the country. Indigenous peoples would not be the ones to lead Peru into a new era, although with the proper technical education they would play their part. It fell upon the European scientific experience, which would take root in Peru through migration, to usher in a new age of greatness, for only when Old World immigrants encountered Peru's "new lands, their virgin forests, their rich and high mountains," would the "ageing European man . . . rejuvenate its life, in the future evolution of history."[34]

The work of García Calderón was influential among his peers, but his knowledge of Peru was disconnected from the national territory, having lived most of his life in France. The same could not be said of José de la Riva-Agüero y Osma, the "father" of the Generación del 900. He traveled extensively through the Andes between 1912 and 1915, carefully recording his experiences, which first appeared in various newspapers and magazines and would be posthumously published in book form. A product of Lima's elite, and one of the few true "aristocrats" of the era, de la Riva-Agüero's travels allowed him to rediscover the sierra. Climbing and descending through the mountain chain, de la Riva-Agüero realized that previous depictions of the Andes did not adhere to its actual realities and lamented, "How did our romantics ignore and falsify the true physiognomy of the Peruvian landscape!"[35]

While de la Riva-Agüero does not mention hydroelectricity, he did explore the Andean rivers that could potentially generate it. Indeed, he seemed to be particularly struck by the force of the waterways. His depiction of the Mantaro and Apurímac rivers is filled with poetry, in contrast with the dry geographic analysis being propagated by the members of the SGL. The Apurímac River, for instance, moved along the "colossal" mountains "with metallic brightness and luster . . . like a great scaly serpent . . . the mythical dragon of the Incas, which roaringly plunges into the mountains of the Antis."[36] Below the rivers were the valleys, which de la Riva-Agüero found in stark contrast to the surrounding mountains, pushing him closer to colonial geographical notions of the Andes. As he was a devout Catholic, this geographic analysis was no doubt informed by his belief that God had created a complex cosmos that was beyond the simple understanding of men.

By the end of his travels, de la Riva-Agüero was ready to speculate on which region would determine the "rhythm" of Peruvian history. The selva undoubtedly held promise, but de la Riva-Agüero foresaw that the exploitation of this region would take hundreds of years. As for the coast, "thin and arid," it could not shoulder the burden of history, leading him to conclude that the future of Peru—as well as its past—was to be found in the Andes. And unlike García Calderón, de la Riva-Agüero considered the Indigenous population to be the main agent of change. Foreign colonists would do great things in Peru—although de la Riva-Agüero correctly predicted that they would never arrive in sufficient numbers—but Peru's Andean peoples could never be replaced, as they were perfectly suited to work in the Andes. He concluded that, regarding the "Indian," Peru "either sinks or it is redeemed with him but cannot abandon him without committing suicide."[37] Although this celebration was born out of resignation rather than optimism, engineers and statesmen in the future would echo the idea that only Peru's Indigenous populations could build complex infrastructures in the Andes.

Republican geography was far from being static during the early days of the Aristocratic Republic. Peru's division into three separate regions fostered myriad debates that understood the complex geographic configuration of the country. Despite dissenting voices, many Peruvian intellectuals did not desire a different geography; rather, they celebrated the republican geography that they had inherited, but they struggled to decipher how to thrive in such a complex geographic and racial environment. In time, hydroelectricity would prove to be the key infrastructure allowing adequate exploration of Andean verticality and use of the perceived abilities of Peru's Indigenous population.

Whether the latter could be fully incorporated into the body politic solely via electricity, however, was still an open question.

AN ELECTRIC CRAZE

The debates surrounding the hydraulic potential of the sierra were likewise informed by the scientific and commercial atmosphere of the time. Foreign scientists arriving in Peru to prospect the country's mineral wealth and teach at new technical educational institutions wrote extensively on how the modern technology of electricity could aid Peru in its industrial endeavors. Unlike earlier travelers such as Raimondi and von Humboldt, broadly defined as naturalists, these scientists were engineers, seeking to find utilitarian purposes for the country's geography. Their work was quickly transmitted by travel books and scientific journals that found receptive audiences in both Peru and the West, as well as by diplomatic officials who presented Peru as a fertile ground for foreign investment. Soon the vast region of the sierra was perceived as uniformly offering an ideal setting for the construction of hydroelectric projects, some of them verging on the fantastic.

One such foreign traveler was the Englishman Charles Reginald Enock, a civil and mining engineer from Birmingham who traveled through Peru in 1904 as a surveyor for American and British mining interests. A fellow of the Royal Geographical Society, Enock blurred the line between travel literature and scientific exploration, and his impressions were collected in the book *The Andes and the Amazon: Life and Travel in Peru*. First published in 1907 the work was popular in Great Britain, where it underwent several editions, as well as being published in the United States. Writing in captivating prose uncharacteristic of an engineer, Enock offered a holistic geographical analysis, although he was careful to note that geographic determinism was not a burden carried solely by Peru. "There is a certain condition of 'geographical continuity,' if I may invent the term, which should be characteristic of the territory of a great nation."[38] Enock clarified that by "geographical continuity" he meant a "feeling of amplitude and extension," which was present in Peru, as the country had access to the Pacific Ocean, the Andes, and the Amazon River basin. The idea of "geographical continuity" was not dissimilar from García Calderón's "harmonious geography," and like the latter, Enock considered that "Peru, almost alone of her neighbors in South America, possesses this characteristic."[39]

Enock traveled mainly through the sierra and the selva, offering his opinion as to which of these two regions held the most promise for the future.

Although in awe of the selva, Enock was particularly taken by the Andes, which he considered to be nothing short of a "mighty hydraulic machine," storing water that produced the streams that flowed down the eastern and western slopes of the mountain chain.[40] The hydraulic machine was to place Peru alongside Western nations and their colonial empires, as hydroelectric stations would power mines, give way to factories, and improve agriculture. This was being done, according to Enock, in disparate places such as Italy, Switzerland, California, and colonial Africa, placing Peru in global imaginaries of development at the turn of the century.

The idea of a hydraulic machine gave the "traditional" mountain chain an allure of modernity, but it was clear that its cogs had been working since ancient times. Engineering may have been a modern occupation, but Enock considered that the Incas had mastered it since time immemorial. There was an inescapable feeling that if ancient Peruvians had been able to thrive in the Andes, there was no reason to believe that modern Peruvians could not do so as well. Furthermore, he had more confidence in the Indigenous population than contemporary Peruvian intellectuals did, who desperately sought European colonists. But Enock saw European immigration as a useful but limited solution. A small number of Europeans might indeed arrive and become superintendents in the region. As for Chinese immigrants, they could survive only in coastal regions. As such, concluded Enock, "Nature has preserved these high regions . . . for the true sons of the soil; those who, at least, have paid her the homage of being born there."[41] Only the descendants of the Incas would prosper in the Andes.

Enock's vision was that the verticality of the Andes had a purpose that modern science was only beginning to decipher, but which remained shrouded in mystery. Despite his engineering background, Enock could not help but feel that, at this moment, the Andes could not be fully understood with the tools available to man. The scientific gave way to the esoteric, as Enock wrote that in time, "science would reveal new and strange" purposes for high altitudes, such as "unknown waves, vibrations, light, dynamic energy, atmospheric products." Yet, while the mysteries of the Andes were still to be discovered, hydroelectric potential was a tangible reality.[42]

Enock's work was well received in Peru, and he wrote articles in Lima's main daily, *El Comercio*, sharing his impressions. He was then contacted by the American Consul in Lima, Alfred L. M. Gottschalk, who forwarded Enock's writings to Washington, where they were read by American businessmen via consular reports. Gottschalk was hoping to entice investment in hydroelectricity by reprinting some of Enock's impressions, including his statement that "the Andes may be considered as a mighty engine, continually intercepting and storing up the moisture of the continent upon its summits, and thence discharging it again under such conditions as create energy in a limitless form and available for the uses of man."[43] All that was needed to tap into this potential was technical know-how and machinery, and industrial development would find a natural home in the Andean mountain chain.

As news of Peru's hydroelectric potential spread abroad, others were popularizing this new technology on the home front. In 1908 Emilio Guarini, a physicist of Italian origin, published *El Porvenir de la industria eléctrica en el Perú*. Widely advertised in the press, Guarini's work was an introductory text meant to familiarize Peruvians with a "new electric age." Electricity, which had already made great strides in Europe and North America, could help countries like Peru, making its agriculture and manufacturing more efficient, as well as modernizing its cities.[44] It could also solve the country's shortage of manpower, a constant concern during the Aristocratic Republic, as the government considered the estimated population of three million to be too low given the country's territorial extension. The electric plow, together with the electric train, could usher in a new era of industrial agriculture despite the lack of workers. This feat would require, according to Guarini's calculations, 10,000 electric plows, 25,000 motorists, and 250 electrical engineers. The possibilities, thus, were endless, and as hydroelectric plants would begin to dot the Andean mountain range, the price of electricity would diminish and profits for exporters rise, pushing the country into a new industrial age.[45]

Perhaps more critically, given the social Darwinist atmosphere of the Aristocratic Republic, electricity could also aid Peru in solving its "racial problem." By mechanizing agriculture, Peru would no longer need to import "undesirable" peoples to work in coastal plantations, such as the Chinese, who further aggravated oligarchic fears of racial deterioration. Guarini considered Chinese immigration as a "necessary evil" given the agricultural needs of the country, but electric machines could eventually put an end to migration from Asia. Entering into some questionable mathematical equations, Guarini claimed the following:

> The Chinese, parsimonious by his weak constitution, as a machine can probably only carry out half of the work that a European adult does on average, that is, a quarter of a horsepower hour instead of half in 8 hours of work. An electric motor that works 24 hours does therefore,

> in practice, the work of 24 by 4 which equals 96 Chinese. An electric plow with a 25-horsepower electric motor can therefore do the work equivalent to 96 times 25 which equals 2,400 Chinese. With only 1,000 electric plows, it is therefore possible to replace 2,400 with 1,000 which equals to 2,400,000 Chinese. The electric plow can therefore, to some extent, solve the problem of immigration in Peru.[46]

Guarini was not mistaken regarding the potential for hydroelectric development in Peru, but at times he proposed unrealistic projects, such as harnessing the power of Lake Titicaca, which led to a dispute with Enock in the press. Guarini, emphasizing his role as a "foreign specialist" who enthusiastically fought for "white coal," was amazed by the potential for energy that the world's highest navigable lake held, considering it even greater than that of Niagara Falls. The generation of electricity in Lake Titicaca would power Peru's railroads, dynamize agricultural and mining activities, and provide the southern city of Puno with first-rate public lighting.[47]

The Titicaca scheme, which received considerable international attention when publicized in American consular reports, was met with skepticism. When interviewed by *El Comercio*, Guarini was confident that his proposal was not only possible but "safe, evident, and not at all complicated." Niagara Falls had a theoretical force of 1,300,000 HP, while Lake Titicaca had 3,800,000 HP, a conservative estimate according to Guarini. The basis for such calculations was far from clear, and figures would change from one specialist to the next. In any case, the problem was that while the Niagara River culminated in a natural fall, Lake Titicaca did not. The project would require the derivation of the lake's waters through the Andes toward the Pacific Ocean. Amazed at Guarini's project, the journalist asked if it was his intention to "drain Lake Titicaca," to which Guarini replied that it would be impossible, for the lake was always collecting water. Regarding the feasibility of the project, Guarini stated that with the advances of modern science it was simply a matter of "time and money."[48] Enock was one of the skeptics. For the potential of Lake Titicaca to become a reality, a complex series of tunnels would have to be built, as well as pumps to elevate the waters through the mountain range to create waterfalls on the western slopes of the Andes. It was far more practical, stated Enock, to use the natural altitude and falls of the rivers near the coast, as some industrialists were beginning to do in Lima.[49]

Such fantastic imaginaries, feasible or not, were useful for propagating this electric potential abroad, as Peru became prominent in the various electric journals that appeared in the West. These journals, the best known examples of which were the *Electrical Review* and *Electric World*, appeared late in the nineteenth century and had the aim of recording "what was happening currently in the industry whether by way of engineering developments or inventions at home and abroad."[50] Charles M. Pepper, an American engineer who wrote extensively on the development of electricity in Peru, stated that "hydraulic and electrical engineers the world over are familiar with the power that the Andes waters hold in reserve." However, he would also agree that some of the "grander" projects would be feasible only after a period of industrial growth. "Such is the proposed tapping of Lake Titicaca, by means of a tunnel through the crest of the Cordilleras, bringing 500,000 HP down to the Pacific. This vast volume of power cannot be utilized profitably for many years, though some day it will be done." There was an air of inevitability regarding the future of the hydroelectric industry, as Pepper would conclude that "the industrial impulse in Peru is in reality too strong to be checked."[51]

Despite such differences, Enock and Guarini agreed that without technical education, Peru's industrial dreams would come to naught. Enock was particularly vocal about this point, emphasizing that the masterful manipulation of waters during its pre-Columbian past would not simply manifest itself in present-day Peru. Peruvian elites believed that for Indigenous populations to thrive they would have to receive a technical education, but Enock was quick to point out that this applied to the very same elites as well. Careers in the army, law, or politics could no longer monopolize the imagination of the upper classes. Guarini was of the same mind, lamenting that "for many centuries, commercial and industrial careers had been considered as inferior to the dignity offered by the arts and the liberal professions," yet celebrated that such a trend was coming to an end as careers in commerce and industry were becoming dominant.[52] The old "Iberian" traditions were being displaced by scientific occupations.

Younger generations—even some members of the oligarchy—were indeed willing to follow a career in engineering, and the state was playing an active if limited role through educational institutions such as the School of Engineers, the National School of Arts and Crafts, and even San Marcos University. These institutions were meant to send their graduates to work for the state as well as the private sector. Enock and Guarini considered these new technical schools to be fine institutions that could train quality engineers, and they were also publicized to foreign readers as symbols of scientific progress during the positivist Aristocratic Republic. In 1912 the School of Engineers—founded in 1876 under the direction of Polish Peruvian engineer Eduardo de Habich and which, in time, would become the National University of Engineering—inaugurated its mechanical-electrical section. The new faculty was led by Peruvian engineer Tómas d'Ornellas and boasted the latest electrical equipment available to students. One of the many English language publications circulating in Peru at the time stated that "the School of Engineers in Peru ranks high among schools of engineering, being considered the best in South America, and in fact, among schools of its class in other countries."[53]

The same publication also celebrated the improved electrical department of the National School of Arts and Crafts, founded in 1905. It was believed that, given Peru's hydroelectric potential, the country needed not only electrical engineers but an army of electrical craftsmen. It was led by Notre Dame University alumni Víctor Arana, who had also taught in the United States and was a member of the American Institute of Electric Engineers. Unlike San Marcos University or the School of Engineers, the National School of Arts and Crafts did not teach mathematics or theory, but devoted time "to training in the shops, to the practical treatment of industrial problems in electricity, and to fit properly the young men, so that, they can earn a living from the very outset."[54] Arana's hope was that such technical education would benefit the Indigenous population, and he proposed that textile industries in the Andes be gradually and systematically mechanized.[55] To this end, Arana presented a project to the government with the aim of creating Indigenous Industrial Commissions. The project was presented to then president Augusto Leguía, who was uninterested in Arana's plans, leading other engineers to lament that "surely the financial caudillo had his head full of banking problems," which diverted his attention from offering Peru's Indigenous population a proper technical education.[56]

When the central state was absent, the prophets of electricity would take matters into their own hands. Guarini, himself a professor at the National School of Arts and Crafts, would take the initiative to popularize the cause of electricity in the provinces by bringing his students on instructional trips, showcasing experiments in distant places such as Ayacucho, Cuzco, and Puno, among other cities. These voyages of instruction were meant to benefit both students and local populations, as conferences were held in public forums in which the benefits of electricity were disseminated. All expenses, pointed out Guarini, were covered by private individuals or by the municipalities, "who have at last understood the true importance of a technical education for the future of Peru."[57]

Despite this enthusiasm, graduates struggled to find adequate work in the nascent electric sector. The lack of job opportunities was not only due to electricity being a new technology but because the Peruvian government insisted on inviting foreign engineers to carry out survey works. Ricardo Tizón y Bueno, future president of the Society of Engineers and the National Society for Industry, criticized the various governments of the Aristocratic Republic, stating that in no other country where technical institutes were sponsored by the state was there a greater dependency on foreigners. What was their purpose if their graduates were not to be hired? The most noted engineers employed by the Peruvian state had been foreigners—Charles Sutton, William Turner, Albert Stiles, and others—who would survey Peru's water sources for the purposes of irrigation, surveys that would also be critical to determining the country's hydroelectric potential. In the past, when no national engineers existed, foreigners had played a pivotal role in establishing the field of engineering in Peru, but, with the ability to create its own professionals, Tizón y Bueno did not understand why Peru still depended on foreign talent.[58] Tizón y Bueno also considered that hiring foreigners went against the very legislation passed by Leguía during his first term in office (1908–1912), that it deemed "convenient for the interests of the country to use in the construction of public works, the greatest possible number of national professionals."[59] If there was a consensus regarding the importance of a technical education, there appeared to be little faith in the engineers produced by Peru's burgeoning higher education system.

Although hydroelectricity had just arrived in the Andes, it was already creating tension as to who should partake in its benefits. Such disputes, however, only reinforced the idea that Peru was destined to embark on a path toward development thanks to the country's fantastic hydraulic potential. Lima's Spanish newspapers raved about the latest developments of the electric industry, and so did papers like *Peru Today* and the *West Coast Leader*, publications destined to publicize Peru to English-speaking readers. Leo Chiozza Money, an Italian economist who had resided in Great Britain since the end of the century, thought that the future of electricity lay in the Andes, and stated that not only should Peru imitate European nations but Europeans should follow Peru's fascination with "white coal." Lamenting that England did not have many sources of waterpower, Chiozza Money would look to the north, to Scotland, where the geographic conditions were not dissimilar to countries that had extensive highlands. "Why is it that what is possible in Switzerland or Bavaria, or even Peru, is impossible for

Scotland?"[60] Although his comment was not without derision—that Great Britain should follow the example of a lowly developed country—Chiozza Money would state that Peru could enjoy that which Great Britain never did: a clean industrial revolution.

A TALE OF TWO CITIES

Despite the electric craze that surrounded the hydroelectric potential of the Andes, much of the country remained in the dark. A study by the U.S. Department of Commerce in 1917 estimated that, "in round numbers," Peru's electric generators had a capacity of 39,000 kilowatts, a far cry from the numbers dreamt by Guarini a decade earlier.[61] In the southern Andes, only large cities such as Arequipa and Cuzco enjoyed the luxury of electricity. The same could be said of Trujillo and a handful of other towns in the north. Huancayo, the largest city in the central Andes, had a small plant that produced 24 KVA, which was used only for public lighting at night, as "the plant is shut down during the day." City authorities hoped to eventually generate 150 KVA, so that "there will be some excess available for small power users, such as bakeries, etc."[62] Other areas that had access to electricity were not urban centers, but mining ones, and they consumed over half of the energy being produced.[63] Such was the case of Cerro de Pasco, where the Cerro de Pasco Corporation (CPC) first built a hydroelectric plant in 1913 on the Yauli River. Outside of Lima, CPC's mining activities represented the only truly integrated electric system in Peru at the time.[64] Life in the rest of the country continued as it had in the past, its rhythm still determined by nature.

Outside urban centers, it fell upon the owners of haciendas or mines to set up small hydroelectric plants. The legal basis for these activities was the Water Code of 1902, which stated that water from rivers was a public good, except for the stretches that passed through private estates.[65] The Mining Code of 1900 was equally liberal, decreeing that mine owners could use rivers for motor force or any other use in the mines.[66] Thus, hacienda and mine owners could exploit the waters passing through their property to light and power their own facilities, or even sell electricity at preferential rates to local authorities to light public buildings.

If one did not own a hacienda or a mine, asking for a water concession was a torturous bureaucratic process. This became even more problematic when Augusto Leguía passed a new water law in 1920. The new legislation required individuals to request permission to use water for motor force purposes. The state would then tax individuals through a complex formula in which an amount would be paid according to the amount of HP generated. Whether the state had the ability to collect the tax was questionable, but, in principle, it would be used for irrigation works as well as surveying waterfalls.[67] Private citizens did not ask for permission en masse, as the state did not make asking for concessions easy. The paperwork was considerable, everything to be submitted in duplicate to the state. Although all calculations had to use the metric system, English measurements were also allowed, for much of the machinery was still imported from Great Britain, showing the absence of a uniform electric system. Once approved, the concession required the signature of none other than the president of the Republic. This would change only in 1938, when the state realized that asking for the president's signature was "inadequate" for the "interests of the state in aiding private initiative." Now all concessions could be approved by ministerial decree, a no doubt useful, but limited, improvement.[68]

The difficulty of asking for a water concession was just one obstacle in transforming Peru's electric craze into reality; the other was the bureaucratic organization of the state itself. The main institution in charge of overseeing Peru's hydroelectric potential was the Ministry of Fomento and Public Works.[69] Founded in 1896 the ministry was meant to represent the positivist ethos of the Aristocratic Republic, as it classified national resources and advocated for the ideal conditions for economic development via the promotion of infrastructure. But Fomento itself was a bureaucratic labyrinth consisting of an aggregate of offices—*directorios*—that acted autonomously from one another, hardly reflecting an adequate division of labor within the state.[70] Beneath the orderly positivist facade of oligarchic government, institutional chaos prevailed.

The case of hydroelectricity illustrated such institutional disarray. When the ministry was created, hydroelectricity fell under the Directorate of Public Works. In 1911, when the Directorate of Waters and Irrigation was established, hydroelectricity was attached to that office. To make matters more complicated, the Directorate of Mines and Oil also had a say when it came to providing hydraulic power to mines. This bureaucratic confusion also applied to the educational institutions sponsored by the Ministry of Fomento and Public Works. For instance, the School of Engineers, which produced not only mining but also electric engineers, fell under the Directorate of Mines and Oil. The National School of Arts and Crafts, another dependency of Fomento, was inexplicably attached to the Directorate of Colonization. Ricardo Tizón y Bueno would express with clear frustration that:

> the directorate of Public Works does not carry out any actual works. Irrigation works are not under its domain, as well as roads, railroads, bridges, and docks that fall under the purview of the Cuerpo de Caminos; drinking water under Hygiene, taking advantage of hydraulic motor force under Mining; and public buildings, in the few cases when they are not under the purview of special juntas, do fall under Public Works, which has merely become a processing office.[71]

The Ministry of Fomento and Public Works not only was a bureaucratic mess but also had institutional limitations. Its presence was felt mostly in Lima, not in the provinces. This meant that the state had scant information about existing hydraulic resources in the Andes. Indeed, the state probably became aware of an isolated hydraulic resource only when a concession was asked. This meant that private initiative was essential to determine Peru's hydroelectric capacity. The first work to truly attempt to do so was Juan Portocarrero's contribution to the study of hydraulic resources for the generation of electricity in 1920. To obtain accurate numbers, Portocarrero had to resort to several sources that reflected the fragmentation within the Peruvian state. The bulletins of the *Cuerpo de Minas* were no doubt useful in this regard. He complemented them with studies carried out by the SGL, as well as data provided by engineers working for the state or for private companies in fields indirectly connected to motor power, such as mining and irrigation. He also used data produced by a handful of meteorological stations located on the coast. The total numbers arrived at by Portocarrero fluctuated wildly. At minimum, Peru had 158,000,000 HP available; at best, 535,000,000 median HP. (Guarini, using figures provided by engineer Eduardo de Habich some years earlier, had arrived at 400,000,000.) Of this, only 43,386 HP were being used, 24,866 for mining purposes and 18,500 for urban centers.[72] Even the numbers had an air of the fantastic during these years.

As late as the 1930s, hydroelectricity remained attached to the Directorate of Waters and Irrigation, yet only one engineer was hired to supervise hydraulic works in the central region, and another for the north and south of the country. The ministry also created a separate entity named the Servicio de Electricidad, which consisted of a skeletal force of two engineers, one administrator, and a typist. Its purpose was to establish public electric services throughout the Republic, but it merely had a regulatory role. To make matters even more complicated, in 1932 the Consejo Superior de Aguas was created, which included the minister of Fomento, the Director of Waters and

Irrigation, and a faculty member from San Marcos University. The Consejo would function as an arbiter if any dispute arose regarding the use of water for various purposes.[73]

Eduardo Lara, who wrote the first history of Fomento in 1936, believed that the ministry had too many responsibilities to be handled by a single entity, and thought that each directorate could form its own ministry.[74] However, it would take many decades for Lara's proposal to become a reality. Change in the promotion of hydroelectricity would come only in the 1940s with the establishment of the Directorate of Industry and Electricity. Fomento itself would be done away with in the late 1960s, when the "revolutionary" government of the armed forces replaced it with half a dozen ministries, including one of Energy and Mines, promoting a greater division of labor within the Peruvian state.

In the meantime, excessive centralization and cumbersome bureaucratic procedures meant that small towns and cities struggled to light their streets. Such was the case in the city of Lircay, a small town located over 3,000 meters above sea level in Huancavelica, the same department where Peru's most ambitious project was to be built half a century later. As early as 1928 a representative for the Ministry of Fomento and Public Works informed the central government that local authorities wished to build a small hydroelectric plant that would use the waters of the Lircay River. The plant required only a 10-meter fall to power a 30 KVA turbine. But local authorities ran into difficulties as the section of the river where the plant was to be constructed flowed through the hacienda Perseverancia, owned by Pablo Vidalón—brother of congressman Damasio Vidalón—as well as lands owned by former congressman Eduardo Larrauri. The state, therefore, had to negotiate with Vidalón and Larrauri for the rights to use the water.[75]

Both Vidalón and Larrauri obtained generous deals. Municipal authorities could use the river and, in exchange, both men would receive free lighting and free light bulbs.[76] The citizens of Lircay waited a full year before the resolution was approved. Another year passed as the turbine arrived at the port of Callao, went through customs, and was transported on mule back through the Andes. By 1930 Lircay still had no light, as transport costs for the turbine had not been paid by the state, and the municipal council had lost interest in the project. A new engineer, Víctor Cáceres, took over the project in 1931. The turbine had finally arrived, but several pieces of machinery had been lost, either during the voyage from Europe, in the port of Callao, or somewhere over the mountains.[77] The paper trail ends that year, and it remains unclear whether the project came to fruition. However, newspaper reports from the 1960s still lamented that Lircay suffered from inadequate public lighting.[78]

If a private individual or a municipal council completed the difficult task of obtaining light, appliances could be obtained only through specialized dealers in Lima, who in turn sold them directly to plant owners. Due to the lack of industrial development, plants in the Andes usually operated only at nighttime, leading the U.S. Department of Commerce to conclude that "the inhabitants of the towns and villages where they are located can be eliminated from the market except for lighting and wiring material."[79] As for the towns that did not enjoy the benefits of electricity, local capital was scarce, and American vendors saw little hope in the development of electric installations in such localities.[80]

Given the suffocating centralization, the electric craze passed by small towns like Lircay, being felt only in large urban centers. Such was the obvious case of Lima, a city that during the first three decades of the twentieth century underwent a process of urban renewal in which electricity played a central part. The transformation of the city was largely the result of President Leguía's vision.[81] After completing his first term as president in 1912, Leguía returned to Peru in 1919 and assumed dictatorial powers. During his eleven-year rule, known as the *Oncenio*, Leguía sought to transform Lima into a modern urban center. This coincided with the demographic expansion of the city. By the time the Oncenio began, Lima's population was 200,000, "ten times as large as when it was founded." [82]

For Santiago Antúnez de Mayolo, later to become Peru's best known engineer, the arrival of electricity symbolized the return of Lima's grandeur, which had been lost since independence. Beyond the boundaries of the "old city," new suburbs were being constructed, large residential houses were being erected, and picturesque boulevards and promenades were changing the nature of the city. Electricity allowed *limeños* to move easily through these new spaces, thanks to modern tramways. It also allowed them to experience the night in new ways, as the introduction of electricity allowed entertainment venues to be open during the night. "Theatre, cinema, and concert performances in the city's cafés and restaurants were extended until twelve or one in the morning, meaning that the public that frequented them could spend more time outside the family environment."[83] In sum, the large rows of electric wiring resting upon wooden posts, Santiago Antúnez de Mayolo argued, had helped Lima lose its air of a "provincial city," which had characterized it since the nineteenth century.[84]

Lima's turning point proved to be the city's celebration of the centennial of the country's independence. As a reporter from the *West Coast Leader* noted during the celebration, at night, the "City of Kings" "seemed transformed as if by magic or by the hand of some enchanter's wand into something really alluring and charming." The city's residents were "set off by thousands of electric incandescent globes, all sparkling and scintillating like the jewels of a Sultan's turban, they seemed like the streets of an Oriental city or a fairy Aladdin's palace."[85] The government palace "stood out in a blaze of light," while the newly inaugurated statue of the liberator José de San Martín was "silhouetted amidst the blaze of the illuminations." The reporter, wishing to lose the crowd, hailed a taxi, which drove him along the "Appian Way" of Lima "under festooned lights over-hung in endless lines building a triumphal arch of national colors outflanked by the two brilliant decorations of several tall edifices clothed in the gorgeous raiment of scintillating lights."[86] Lima was indeed no longer the "provincial city" that Antúnez de Mayolo had once lamented.

But Lima's rise as a modern city was also viewed with suspicion by some of the most prominent members of the city's elite, who experienced what Wolfgang Schivelbusch has described as a "disenchanted night."[87] Although far from smashing the city's lanterns as the Parisians of the nineteenth century did, José Gálvez—one of the foremost poets of the time and a member of the Generación del 900—chronicled how popular customs and superstitions that had been commonplace in the city were disappearing in a nostalgic ode to an "old" Lima that was fading away.[88] Such was the case of the city's ghosts and spirits, which seemed to vanish with the arrival of electric light. "One of the manifestations of the *other life* which had the most terrifying meaning occurred when the lights would unexpectedly go out." The arrival of electricity meant the disappearance of candlelight, and hence the disappearance of ghosts, who had previously manifested themselves when the winds blew through the house, for "ghosts today have yet to find a way of turning the electric switch."[89] But the disappearance of the chaotic and mysterious night was replaced by new superstitions from Lima's popular classes, particularly those regarding the generation of electricity itself, as Lima's poorest citizens remained suspicious of its origins, at times thinking that it represented the work of dark forces. Quoting a *zambo*—a colonial-era term to describe a person of Indigenous and African ancestry—who was incredulous that the light was coming from the waterfalls in the distant sierra, the following line says much about how electricity was received in the capital: "Pa su aguela.

FIG. 3. Peru's Centennial Celebrations in the city of Lima. Antúnez de Mayolo, Santiago. *La génesis de los servicios eléctricos de Lima* (Lima: Impr. E.Z. Casanova, 1929). Biblioteca Nacional del Perú.

That the gringos have made light from the stuff that puts out candles, that is the work of the devil!"[90]

Electricity may have evicted ghosts and spirits from the city limits, but it had yet to fully revolutionize everyday life, as those who did welcome the arrival of electricity saw in it potential benefits for maintaining the existing social order. Víctor Arana, who had enthusiastically proposed bringing electricity to Peru's Indians, was more skeptical about the impact of the new technology on women. He suggested that it ought to be limited to the domestic sphere, an idea he propagated in his manual *La electricidad en el hogar*. Just as electricity could usher in a new era of modernity, it could also reinforce and, indeed, save tradition in times of social change. Electricity could help rescue traditional home life and even reaffirm it with greater strength than before. Electricity, thus, could not only solve Peru's "Indian question" but also its "domestic problem," represented by the nascent feminism of the time. For Arana, electricity could give way to a "conservative feminism." If women "fell in love" with the electrical devices and machines that would aid them in carrying out thankless domestic tasks, "would they not have a greater personal satisfaction, more intense and longer lasting than that of carrying out their duties in the offices in which they work?"[91] Arana believed that a professional career could never compete "with the noble and feminine tasks of the home, made noble and purer by the Spirit of Electricity." Feminism was to be found within the home, not outside of it. This was the goal of electricity in the domestic sphere, to "feminize feminism, for if taken to the extreme, this social movement could masculinize women."[92] New histories of energy have highlighted how it was met with fear in the West, as it could not only cause physical harm but also threaten the existing domestic order.[93] In the case of Peru, it seems, electricity could make sure that said domestic order did not change.

Arana proved to be too optimistic. As late as 1927 the U.S. Department of Commerce considered Peru to be a fertile market for electric equipment, but not so for household appliances. "Cheap household labor, the types of dwellings, and the cost of electricity all operate to restrict the use of appliances to a minimum."[94] Two years later, it was estimated that "4,000 electric appliances are used in Lima, Callao and suburbs, of which electric fans, irons and grills share first place with about 1,000 each; toasters, kettles and percolators about 250 each; and other appliances in lesser numbers."[95] The purchase of electric appliances was limited by the lifestyle of limeños. Antúnez de Mayolo commented that electric kitchens were expensive because of the electric tariffs that limeños would incur when preparing their favorite dishes. "The type of diet is also of great importance, for in Lima it is customary to prepare soups, stews and broths . . . all which require long cooking times and much heat."[96] Likewise, the city's residents did not consider other appliances necessary. Used to cold but bearable winters, "the residents prefer to become

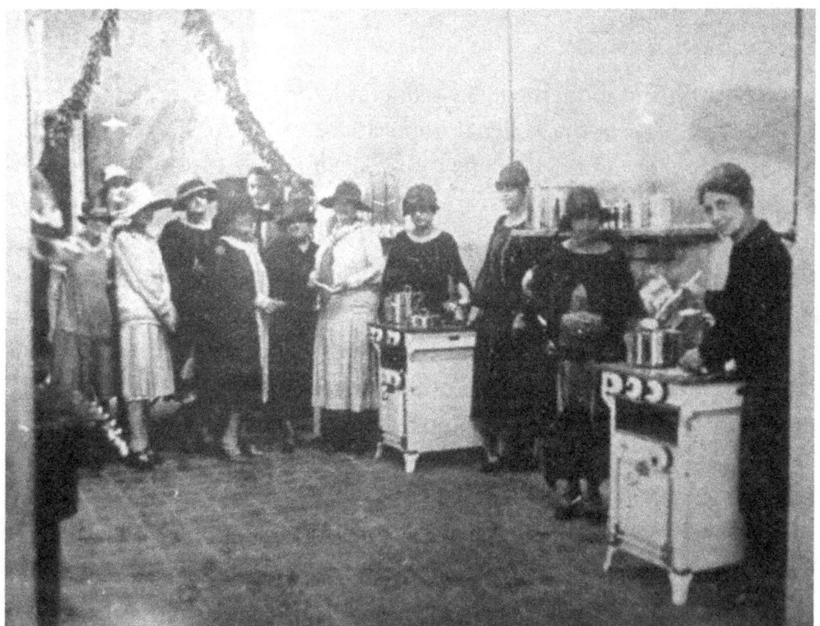

FIG. 4. "Display of two electric kitchens at the 'Bien del Hogar Society.'" Antúnez de Mayolo, Santiago. *La génesis de los servicios eléctricos de Lima* (Lima: Impr. E.Z. Casanova, 1929). Biblioteca Nacional del Perú.

accustomed to the chill of winter rather than have artificial heat." A further obstacle was presented by the fashion tastes of Lima's Europeanized upper class, as they considered appliances from the Old World to be better suited to their aesthetic needs. "Lately a few chandeliers have been brought in from the United States, but the style of the European chandeliers appeals more to this market . . . chandeliers from the United States . . . are not fancy enough to suit the taste of the purchasers, who have been accustomed to the more ornamental products of European factories."[97]

Lima's residents imagined electricity in different ways, some receiving this modern technology with awe, others with concern, but there was no denying that this new technology was transforming the old colonial city. However, by the end of the Oncenio, a new generation of intellectuals, uninterested in the nostalgic concerns of Lima's conservative oligarchs, began to question why Lima alone should enjoy the benefits of electricity, seeking a more equitable division of the hydraulic spoils of the sierra among the provinces.

A NEW GEOGRAPHY

Lima's growth and modernization—and the lack thereof in the provinces—sparked a debate on the political implications of the country's geography. A new generation of intellectuals—radically different in outlook from the positivist Generación del 900—attacked Lima's political dominance, particularly as it prospered at the expense of the hydraulic machine that was the sierra. Unlike the old guard, these new intellectuals were from the provinces and championed political and economic decentralization. They were also representatives of or sympathized with the social and political movement known as *indigenismo*, a movement that celebrated Peru's Indigenous populations, as opposed to the oligarchic view that considered them the source of the nation's ills.

This tension between capital and province was best expressed by Moquegua-born José Carlos Mariátegui, an unorthodox Marxist who published his famous *Siete ensayos de interpretación de la realidad peruana* in 1928. Mariátegui conceded that while Lima was experiencing great urban growth, it was nevertheless destined to fade as a great city, as its growth was an illusion. "The spectacle of the development of Lima in recent years moves our impressionable limeños to deliriously optimistic predictions about the future of the capital. The new suburbs and the asphalt avenues . . . persuade a limeño—under his skin-deep and cheerful skepticism—that Lima is not far behind Buenos Aires and Rio de Janeiro."[98] But for Mariátegui, Lima's position was the result of the city exploiting a centralist regime; its primacy was based on political or administrative privileges, not on economic ones determined by geographical advantages.

Mariátegui did not break from the established tripartite division of the country; rather, he used it to argue that Peru's diverse geography meant that no political centralism could be successful in the long run, for "the physical geography of Peru runs counter to centralization."[99] As a Marxist, Mariátegui was not interested in the formal aspects of decentralization—a political or administrative process—but rather in its economic implications. Through this materialist lens, Mariátegui argued that Lima's dominance had not always been assured. In colonial times, Spanish lust for gold and silver meant that the sierra was the economic center of the viceroyalty. It was only during the guano boom that the power of the coast was strengthened, accentuating "the dualism and conflict which now constitute our greatest historic problem."[100]

However, after the War of the Pacific, Mariátegui saw little merit in Lima's claim to dominance.

Mariátegui's fight against centralism was informed by the aftermath of the Bolshevik Revolution. Mariátegui saw industrial development as inevitably tied to electrification and was mesmerized with the Soviet GOELRO plan ("State Commission for Electrification of Russia"). "Russia, the motherland of a burgeoning socialist civilization, works feverishly to develop its industry. Lenin dreamed of the day his entire country would have electric power."[101] Through his *Amauta* magazine, Mariátegui and his followers would publicize the accomplishments of Soviet five-year plans, especially the ongoing construction of the Dnieper Hydroelectric Station, and, much like Lenin, thought that socialism was an unattainable goal without the development of electricity, a necessary requirement for industry.[102]

Informed by the Russian experience, Mariátegui argued that Lima's lack of industrial development cast doubt on the city's claim to being the political center of the country. After all, Mariátegui claimed, cities such as London, New York, Berlin, and Paris owed their rightful place as political centers to their industrial importance.[103] If Lima were to become a great industrial center, no one would question its right to exert its political power. But Lima's geography meant that its industrial development looked bleak. Industry manifested itself in the great capitals because they tended to be the center of a country's transportation network or because they had access to coal and iron, two requirements that Lima did not meet. If no coal was available, industrial centers had to be close to other sources that generated electricity. "In these days of worldwide electric power, a third factor that attracts industry to a site is the proximity of hydraulic resources. 'White coal' can work the same miracles as black coal to create industry and cities. Lima has none of these factors; its surroundings do not attract industry."[104] Lima was destined to fade away as a city of any importance, for the territory that surrounded it was not promising for industrial development. Time, however, would prove Mariátegui wrong on this point, at least with regard to hydroelectric power.

Mariátegui was not hopeful about the prospects for decentralization. He considered that any devolution of power to the provinces would strengthen the position of Peru's landowners and reinforce the exploitation of Peru's Indigenous populations. In true *indigenista* fashion, Mariátegui thought that real decentralization would occur only if Peru's Indigenous population was freed from the cruel grip of landowners, unleashing their socialist tendencies.

This *indigenismo* had a great impact on regional intellectuals, who by the 1930s were clamoring for greater decentralization. Among this group of regional intellectuals was a new generation of geographers who joined the SGL. One such geographer was Emilio Romero Padilla, who took geography to its natural political conclusion, aiming for a total reconfiguration of Peru's political and economic geography. Born in Puno in 1899, Romero moved to Arequipa to study at the National University of San Agustín. Eventually, he moved to Lima to continue his studies at San Marcos University. It was during these years that, like most *provincianos* arriving at an unwelcoming capital, a young Romero went to Mariátegui's house, where he was interrogated about Puno's economic and social conditions. His career would progress in the coming years. He became a professor at San Marcos University and eventually held the presidency of the SGL, from 1945 to 1948 and then again from 1958 to 1979. Interested in both economics and geography, Romero became a national pioneer in the discipline of economic geography, highlighting the connections between Peru's topography, national development, and political configuration. For Romero, economic geography meant not only studying the activities carried out by man that are indispensable to his subsistence, but also the use of natural resources required to achieve the greatest possible economic development of the nation.[105]

His provincial background and close affinity with indigenista trends allowed him to have a particular view of Peruvian geography, which was different from that of Lima intellectuals in the SGL.[106] In his landmark work, *Geografía económica del Perú*, first published in 1930 and later adapted into multiple revised editions, Romero celebrated Peru's geography. Romero conceded that the Andes were "a gigantic barrier that reaches the most sublime heights," and that it was the main factor that molded the economic geography of Peru. Despite this, Romero stated the Andes made Peru a remarkable country, giving it the most complete and original geography on the globe, as well as being the cradle of the largest pre-Columbian civilization in the Americas. As a man born in the highest parts of the Andes—the *puna* or *altiplano* of southern Peru—for Romero, the Inca Empire and its predecessors could not have emerged in a different geographic context.

Following Mariátegui's dictum that Peru's geography was incompatible with centralization, it was perhaps inevitable that Romero's analysis of Peru's economic geography would venture into the realm of political geography. In the opening pages of *Geografía económica del Perú*, Romero states that the "old" geography, which had merely observed, had come to an end, and that any new Peruvian geography must dialogue with the emerging trends of a broader human geography, including its political aspects. While respectful of the naturalists of the past, such as Raimondi and the early members of the SGL, Romero considered their achievements to be limited to the study of physical geography, which had administrative ends rather than political ones.[107] A political vision of the country, in the days after the First World War, should be based on current and future economic potentials, so that Peru could adequately interact with other states. This potential, according to Romero, was stifled by Peru's existing political configuration.

Mixing politics with geography, Romero became a founding member of and congressman for the Decentralist Party. In 1932 he would publish *El descentralismo*, outlining his proposal to use geographical knowledge to rearrange Peru's political divisions.[108] Twelve departments would be created based on variables such as availability of natural resources and distance between population centers. Romero argued that Republican Peru had never enjoyed political cohesion, and that only the Incas had given the country a political and administrative organization that was adequate to its geography. Even the Spanish, with their horses, organized the colonial system according to distance, as was reflected by the parishes and bishoprics that were meant to convert the Indigenous population. Thus, both the Inca and the Spanish empires had organized Peru's territory in a logical manner, "upon a geographical basis." Republican Peru had maintained previous organizational demarcations but had failed to update them to the necessities of modern economic and political life.[109]

Distance and resources meant that Romero's proposal was expressed in infrastructural terms: new departments ought to be created according to the opportunities for building roads and irrigation.[110] Yet, while hydroelectricity receives scant attention in *El descentralismo*, Romero does include it in his earlier *Geografía económica del Perú*, already linking the new technology to his ideas about decentralization. Romero stressed once more that Peruvian geography was exceptional not because of the challenges that it presented, but because of the possibilities that it offered. One of these possibilities was the generation of hydraulic energy, an offering very much linked to the constant evolution of technology. Hydraulic energy was particularly important not only because, unlike coal and oil, it was a renewable resource but also because the very geographic makeup of the country made its exploitation more advantageous and economical, for "Andean topography gives an energy value irreplaceable to the rivers of the Andes." For this very reason it was possible to see many localities, "lost in the most rugged Andean corners, under the freezing fog of the mountain peaks, enjoying electric energy."[111]

Romero would conclude that on paper, Peru's hydraulic resources were less than those of Argentina, Brazil, or Mexico, yet its hydraulic potential was greater "because of the very shape of our soil, for eminently geographic conditions." In the borderlands between the coast and the sierra, there were no great navigable waterways; instead, the region enjoyed small but resounding rivers. Another positive attribute was that nature had wisely dispersed these hydraulic sources throughout the country. "Instead of giving us one immense waterfall . . . capable of producing immense energy, nature has distributed rivers and streams throughout the whole of our territory."[112] The very topography of the country dictated political decentralism, and hydraulic energy was the most decentralized resource available in Peru.

The "new geography" elaborated by regional intellectuals managed to permeate, to some extent, oligarchic views of Peru's geography. Such influence found best expression in the work of José Pareja Paz Soldán. Pareja Paz Soldán was not a professional geographer but a well-known constitutionalist and diplomat from one of Lima's most prominent families. Diplomats were expected to have knowledge of a variety of subjects and, given the importance of geography in both national discourse and international relations, Pareja Paz Soldán could not have been expected to excel without a profound knowledge of it. His intellectual position was best expressed in the lecture notes for his introductory geography course at the Pontifical Catholic University of Peru in the 1930s. Citing equally de la Riva-Agüero and Mariátegui, these notes served as a basis for his two-volume work simply titled *Geografía del Perú* and a manual of the same name, both published in the 1950s.

Although respectful of the "great socialist thinker" Mariátegui, Pareja Paz Soldán openly denounced the "materialist" indigenista movement. This was partly because of his staunch Catholicism and opposition to Marxism. But he also considered it to be antipatriotic, as it put the country between two extreme positions, the Western or European and the indigenista.[113] Rather than having to choose between the two positions, Pareja Paz Soldán proposed his own notion of an "integral Peru" characterized by *mestizaje*, in which both Indians and Spaniards had been united. Furthermore, each group contributed to the national character. From the Incas, Peru had learned to master nature, to conquer distance, and the collaboration between citizen and state. From Spain Peru had inherited Western values and philosophy, best expressed by the Spanish language and Catholicism.[114]

He noted, however, that this "integral Peru" did not yet constitute a nation, in part because of a racial obstacle ("whites, Indians, blacks and mestizos, we all live in motley and confused promiscuity"). Yet Pareja Paz Soldán also considered previous characterizations of Indigenous peoples as inferior to be mistaken, for the criteria for measuring the virtue of a race was its capacity to thrive in its environment, a test most successfully passed by the Andean people.[115] Despite this celebration, his analysis was reminiscent of that of the Generación del 900. With a clear social Darwinist influence, he noted that the Indigenous population shared the physical characteristics of the Andes, which were "dull and grey." Such monotony had produced a "quiet and passive" people, who, unlike in earlier periods, were unwilling to "wrestle with nature and snatch her secrets."[116] Even the positive attribute of surviving in high altitudes because of their different biological makeup—an idea already being popularized by Carlos Monge, who characterized Peru's Andean populations as "men of the altitudes"—was a problem, for when someone from the sierra migrated to Lima and found some comfort in the capital, in a fit of sadness they would suddenly "return to their primitive Andes."[117] His conclusion was damning, overall, and no different from what García Calderón had asserted at the beginning of the century: that while not "unadaptable to civilized life" Indigenous peoples at best would prove "useful."[118]

The racial problem was of great importance, but the main obstacle to the creation of a nation was the absence of a "geographic unity." In part, this was the result of Peru's geography, in part the result of its political configuration. He certainly believed that decentralist arguments had validity. But unlike Mariátegui, he did not dispute Lima's claim to being the capital and thought the obsession of decentralist intellectuals with the city distracted them from the country's "defective" political demarcation, which had indeed challenged "the development of provincial life" through the excesses of a centralism that "paralyzes regional activity."[119] If such centralism could be overcome—without Marxism or negatively affecting Lima—the Andes would have much to offer.

As for the topography of the Andes, Pareja Paz Soldán described it as a paradox, a place characterized by both challenge and opportunity, a dynamic best represented by the potential for electric generation. "Peru is a country of mountains and buttresses. These two accidents make difficult the creation of lines of communications, keep populations apart and increase the transport costs of agricultural products and manufactures." But these difficulties were in part compensated for by the vast amount of mineral wealth that was to be found in the region, as well as by its rivers. Indeed, it was the "Andean river" that once again showed potential, for while they had no geographic purpose in the highland ("they have neither the unifying function that they perform in the rainforest nor the fertilizing one on the coast") they did represent the "possibility of creating hydroelectric power at a low cost, by taking advantage of the marked water inclines that exist throughout the territory, which if efficiently captured and used will become powerful sources of national wealth."[120]

For Pareja Paz Soldán, the Andes held the key to Peru's progress, for the "production of energy is the basis of all industrial development." Only countries that enjoyed such sources of energy and could produce it at low prices could expect to become manufacturing and mining powerhouses. Hydroelectric energy was particularly advantageous as it was cheap, easy to transport, and left few residues. Indeed, although not citing any sources, Pareja Paz Soldán finds a rare opportunity to place Peru among the ranks of developed countries based on the potential of electric generation. Such potential was 0.05 HP per habitant, placing Peru fifth in the world, behind Canada, Switzerland, Italy, and the United States.[121] By situating Peru alongside developed countries, Pareja Paz Soldán hints that the Peruvian Andes, while presenting a challenge to nationhood and social progress, also offered great potential for considerable industrial development. What remained was to turn this potential into reality.

Much like their predecessors, Romero and Pareja Paz Soldán agreed that the Andes were a hydraulic machine that could push Peru toward industrial development. It remained unclear, however, if the population of the sierra offered the same potential as the very topography of the region. Romero, the indigenista, certainly believed so, while Pareja Paz Soldán—writing at a time when the great waves of migration were no more—was somewhat less generous yet not entirely pessimistic regarding the "men of the altitudes." Regarding decentralization, old divisions remained between provincial and Lima geographers. Despite this, this new generation of Peruvian intellectuals, much like their predecessors, saw a complex geography with possibilities, offered by the arrival of new infrastructures, that were yet to be deciphered.

CONCLUSION

By the early twentieth century, the division of Peru into three natural zones had become an inescapable reality, although that reality had been created by geographers. However, within the first three decades of the century, there was a dynamic debate as to the virtues of Peru's fragmented geography. In this debate even the most conservative voices admitted that the hydraulic potential of the Andes offset some of the problems that it presented for the development of other forms of infrastructure. The debate remained as to what role Peru's Indigenous population would play in this ambitious endeavor. There was little agreement on this point. Either Peru's Indigenous populations would be the only ones able to thrive in the Andes, by virtue of their particular strengths, or, as García Calderón stated, they would merely be useful if incorporated into this new scientific order through the acquisition of a technical education. By 1931 hydroelectric debates and imaginaries allowed seeing Peru, in the words of intellectual Jorge Basadre, "geographically and economically, as a problem and a possibility."[122]

Many of these aspirations were of an abstract nature, as hydroelectric plants were slowly built, and thus imaginaries remained on paper. It would take some time for the Peruvian state to transform promises into reality and include hydroelectric infrastructures as part of its political mission, as its institutional configuration remained chaotic and its presence in the national territory limited. Rather than promote the right conditions for development, the improvised bureaucratic structure of the Ministry of Fomento and Public Works hampered the development of Peru's hydraulic resources. Direct state intervention was avoided, but private initiative was likewise stifled by excessive bureaucracy. Peruvian elites might have been seduced by the promise of infrastructure, but they were also unwilling to develop a state apparatus that would adequately take advantage of it. Under such conditions, it is not surprising that the true birth of Peru's hydroelectric industry would emerge from the most prosperous elements of civil society and, in time, be appropriated by the state.[123]

In the meantime, unfulfilled promises resulted in a quick politization of hydroelectric infrastructures, or at least its imaginaries. Since the Andes was the "hydraulic machine" that would make possible the dreams of electrification, regional intellectuals resented that the benefits of electricity were being disproportionately felt in the capital city of Lima. The political promise that was decentralization was given legitimacy by infrastructural arguments in general, and hydroelectricity in particular. Indeed, by the end of the Oncenio, Mariátegui asked, "Lima is the capital today, but will it be the capital tomorrow?"[124] Yet despite the push for greater political equality between Lima and the provinces, Lima would strengthen its position, in part thanks to the development of hydroelectric infrastructures by exploiting sources of "white coal" near the capital. Thus, the linkages between infrastructure and politics would only intensify as Lima's elites began to push the electrical frontier eastward, promising an electric development that was to benefit both the capital and the interior. Regional elites, however, feared that such promises would remain unfulfilled.

Vertical Limits

2

On May 1, 1908, Augusto Durand, coca baron and leader of the Liberal Party, boarded the train to the town of Chosica from Lima's central station. He was accompanied by several followers armed with Mauser rifles, as he had hatched a plot to oust newly elected president Augusto Leguía. Arriving at Chosica, Durand, dressed in black and with a slouch hat of the same color, stepped out onto the platform and stated, "Those who are not with the revolution, leave now!" He then headed toward the offices of the Lima Light and Power Company (LL&P) and asked for the power produced by the Chosica power plant to be cut.[1] Lima did suffer a partial blackout, but Durand's insurrection came to naught. *El Comercio* reported that Durand's actions were meant "to create a surprise, the only way of achieving a revolutionary movement in these times . . . that is why Dr. Durand cut off the electric current from Chosica to Lima, because by generating darkness here, there was a greater possibility of creating the confusion needed for surprise." Other sources stated that cutting off the power was essential for Durand to reach Cerro de Pasco, with "the hopes of finding support in that rich province," or simply to "seek refuge in the jungles of Huánuco, if he is already alone and abandoned."[2] Despite its failure, Durand's attempted insurrection symbolized the first time that LL&P—the country's largest electric company—became embroiled in Peru's political disputes.

This chapter traces the history of LL&P from its foundation, early in the twentieth century, to the end of the Second World War. During these years, LL&P erected a complex hydroelectric system that exploited the verticality of the central Andes, constructing eight hydroelectric plants in a stepped fashion. This electric pyramid—financed with Italian and Swiss capital—was gradually built in accordance with the city's growing population. While the company proved formidable at overcoming geographical difficulties through clever engineering techniques, it proved less adept at dealing with the political challenges that emerged during the interwar years.

While the company emerged in a context of laissez-faire economics, the transnational nature of its infrastructural investment meant that LL&P's financial and engineering activities could not be isolated from the volatile politics of the time. On the home front LL&P, led by Italian administrators that espoused the ideology of fascism, clashed with its workers, who, during the Great Depression, became unionized under the leadership of Marxist-inspired groups. Domestic political struggles were enhanced by the global political imaginaries of electricity, bringing to the fore the ideological dimensions of infrastructure.[3] Hydroelectricity could be associated with the importance of public works in fascist Italy, Marxist ideas regarding industrialization and revolution in the Soviet Union, or the liberal experience of New Deal America. Thus, while vertical in geographical terms, hydroelectricity worked horizontally across the political spectrum.

The study of the impact of infrastructural development on Peruvian politics during this period has focused on the domestic setting, portraying it as a tool employed by the state to control its population and weaken political opponents.[4] But the development of hydroelectricity also had profound international ramifications. The totalitarian aspirations of LL&P's Italian administrators would eventually be challenged by a pro-Allied government during the Second World War. Hoping to counter the influence of fascism in the capital—including LL&P and its Italian linkages—the Peruvian government promoted hydroelectric projects that involved former engineers of the Tennessee Valley Authority (TVA). Such efforts were pursued not only to satisfy the United States but also to promote national ideas of modernization and development, highlighting the agency of the Peruvian state during the war years and adding a new layer of complexity to Peruvian-U.S. relations.[5] Such development also allows infrastructure to be factored in as a variable in Peru's foreign policy, complementing existing diplomatic approaches.[6]

The early history of LL&P also sheds light on how transnational financial and scientific networks expanded during this period. While Italian administrators became public figures, Swiss financiers silently poured capital into the company, underlining the importance of global financial connections in the development of the capital-intensive electricity sector.[7] The establishment of financial networks also determined the itineraries of engineers, as the Swiss Pablo Boner surveyed the Andes and built plants that were deemed to be highly sophisticated by international standards, providing an early example of "scientific excellence from the periphery."[8] Local figures also played a critical part in the process, informing the company of the location of water sources. This proved that local knowledge was essential for the successful application of Western technologies.[9] In the end, Swiss interests took over the company, particularly as Peru's political battles frustrated LL&P's fascist managers, who were unable to order and control Peruvian society as the company had done with the surrounding Andean environment.

BLUE DEVILS IN THE MOUNTAINS

The modest origins of Peru's largest electric company are to be found in the activities of Lima's small industrialists in the second half of the nineteenth century. To power their fledging factories, small plants were established to take advantage of the city's scattered canals and streams. Such was the case of Bartolomeo Boggio, an Italian businessman who arrived in Peru in the 1870s. Like many Italian immigrants during the period, Boggio was originally devoted to commercial activities, but he made incursions into the textile industry after accumulating capital from the wool trade.[10] In the late 1880s Boggio established the Santa Catalina Textile Factory, powered by a small hydroelectric plant on the Huatica River in downtown Lima.[11] Boggio would then form a partnership with Juan Manuel Peña Costa and Mariano Prado Ugarteche—the latter a member of one of Lima's most prominent families—renaming the enterprise the Santa Catalina Industrial Society.

As the factory expanded its activities, it required more energy. A separate entity, the Empresa Transmisora de Fuerza Eléctrica, was formed in 1885, and it acquired an importance of its own as it began to supply power to nearby factories. New businessmen Pedro Ugarteche and Guillermo Espantoso also came onboard.[12] A new hybrid plant—half hydraulic, half thermal—was inaugurated on the left bank of the Rimac River, and the Santa Rosa Electric Company was created in 1900.[13] By then the company was wholly controlled by the Prado family, who had bought all the company's shares, and Mariano Prado became its manager. Electricity would thus serve as one of the foundations of what was to become the "Prado Empire," symbolizing the industrial and financial importance of one of Peru's main oligarchic families.[14]

The case of the Santa Catalina Electric Company was by no means unique. As small industries sprang up in different parts of the city, they too developed their own sources of motor power. Such was the case of the Sociedad de Alumbrado Eléctrico y Fuerza Motriz de Piedra Liza, founded by the Spaniard Juan Peral, and the Compañía Eléctrica del Callao, established by another Italian, Faustino Piaggio.[15] As the fortunes of the Prado family grew, smaller companies were quickly absorbed. The final step taken by Prado to establish control over all of Lima's electric services was to absorb the city's tramways. The merger between the two industries was possible because Lima's tramways—previously pulled by horses—now required electric power. There was also considerable endogamy between the two activities, as the Prado family already owned shares in the tramway business. Between 1906 and 1910 all the companies were united, in the form of a trust, under the name Empresas Eléctricas Asociadas (EE. EE. AA.). Each company maintained its own legal entity when executing contracts in its respective areas of concession but was able to cut costs by having a single administrative organism and sharing technical services.[16] EE. EE. AA enjoyed a natural monopoly in the city, and due to the laissez-faire nature of the Aristocratic Republic, it would not face any public competition.

The newly created electric trust had social capital of 1,500,000 Peruvian pounds, a sum hailed as "without precedent in the industrial history of our country."[17] This accumulation of capital was made possible because the company brought together the most successful economic groups of the era. The Prado family, representing the most powerful non-agrarian consortium of the time, had access to capital provided by the Banco Popular del Perú. The second group consisted of Italian financiers who were commandeered by Gio Batta Isola and who had connections with the newly established Banco Italiano. The last group consisted of the Peru and London Bank, represented by the larger-than-life figure of José Payán, a Cuban national who, after abandoning his career as a doctor and being exiled for his participation in Cuba's struggle for independence, migrated to Peru, becoming a respected economist and financier. The electric industry, hence, represented the more progressive and ambitious elements of the Peruvian oligarchy, whose main interests were to be found not in the ownership of land but in financial and industrial activities.[18]

From its very formation, the company had strong transnational linkages. This trend was reinforced the same year as the merger, as EE. EE. AA. obtained a loan from the Henry Schroder & Co. Investment Bank. The Schroders—an old Hanseatic family that had resettled in London in the mid-nineteenth century—had been active in Peru since the days of the guano boom but had abstained from further investment even before the country's disastrous defeat in the War of the Pacific. The first company to again receive a loan from the Schroders was EE. EE. AA., placing it at the forefront of the economic resurgence that characterized the Aristocratic Republic. The economic ties between EE. EE. AA. and the Schroder Bank remained active throughout the company's history, and the Schroders, particularly through their New York office, helped the company by placing shares in the American market when further expansion was required.[19] The loan from the Schroders also gave the

FIG. 5. Offices of Lima Light and Power Company, downtown Lima. Empresas Eléctricas Asociadas, *Souvenir of the Empresas Eléctricas Asociadas (Lima Light, Power and Tramways Co.)* (Lima: M. Moral, 1913). Pontifica Universidad Católica del Perú, Ferreyra Collection.

company considerable international cachet, as it required the establishment of a British office to represent the interests of its new shareholders. The company adopted the English name Lima Light, Power and Tramways Company, to be used in all official international transactions. Limeños, however, would simply call the company "*las Eléctricas*."[20]

The Santa Rosa Electric Company was able to lead the merger not only because of the Prado family's capital but because it proved more adept at embracing technological innovation. Santa Rosa was the first company in the city to establish a hydroelectric plant on the outskirts of the capital, in nearby Chosica. Known as the "Nice" of Lima, sunny Chosica was a quiet locality where some of Lima's most affluent families established summer homes to escape the capital's humid winter. Located at almost 1,000 meters above sea level, it also symbolized the end of the coast and the beginning of the Andes. Between 1903 and 1907 two hydraulic plants—Chosica and Yanacoto—were established not far from the quaint center of town, allowing Santa Rosa to outproduce its competitors.[21]

Both plants, which produced a combined energy output of no more than 8,000 KW, caught the attention of both national and foreign engineers, who highlighted the innovative nature of placing the plants at different altitudinal levels. This enabled the Yanacoto plant to reuse the waters that passed through the Chosica power station, harnessing the verticality of Andean rivers and thus making the construction of large dams unnecessary. Noted engineer Charles Melville Pepper, in his broader study of "Electricity in Peru"—published in the prestigious British journal *The Electrical Review*—considered Chosica and Yanacoto to be the best examples of hydraulic engineering on the continent. Pepper's description of both plants shows the careful modification of water routes and clever use of altitudinal levels to produce electricity, challenges that the company would apply on a larger scale in the 1930s. The Chosica plant built a canal to divert waters from the Santa Eulalia River into the Rimac River. At the point the two rivers became one, a small concrete dam was built, diverting the water into a kilometer-long canal that led into two steel pipes, "one and one-half meters diameter each and 670 meters long."[22] The effective "head"—the change in water levels between the point where it enters and where it exits—was almost 50 meters. As the water violently reached the station, it hit the wheels of the turbines at a velocity of 1,800 meters per minute, producing roughly 3,000 KW.[23]

Once the water passed through the Chosica Station, it was redirected to the Yanacoto plant. Before it passed through this second stage, steel gates at different levels would "remove the immense quantities of sand and rocks that the river carries during high water."[24] The canal was 5 kilometers long, most of it "being in earth," the remainder consisting of cement walls. Once the water reached Yanacoto, it fell 80 meters, producing almost 5,000 KW.[25] Both plants were equipped with sophisticated technology. In time, Swiss technology would be dominant, but for the time being LL&P depended on British and American machinery, and both plants boasted British Pelton double-wheel turbines and General Electric generators.[26]

Engineers celebrated not only the clever redirection of water routes and the precise positioning of both plants but also the problems faced when connecting them to the capital. A. L. Kenyon, an American hydraulic engineer working for LL&P, noted the challenges that "the first energy transmission system in South America" faced when descending from the Andes into lower altitudes. Kenyon was particularly concerned with the impact of humidity on the 40-kilometer-long connection between the Chosica plant and LL&P's distribution center in Lima. "It might seem that the western coast of Peru, with an entire absence of rain and lightning, offers to the engineer an inviting field for high-tension work. On the contrary . . . the heavy, damp fogs that drift up from the sea during the winter season penetrate to every part of the insulator and pin." Lima's infamous humidity would bring a "thick coating of muddy paste that only periodical washing by hand can remove." The problem had been caused, it seems, by "misplaced confidence in the climatic conditions . . . and the insulator selected." Glass insulators would frequently break and crack the wooden poles, creating "a hissing noise made by the leaking current (that) could be heard from two to three poles away." The fires and hissing sounds would frighten the "superstitious natives," who at night claimed to see "blue devils" in the mountains. It seems that using porcelain insulators eventually fixed the problem, effectively exorcising the "blue devils."[27]

Scientific pilgrimages to Chosica became commonplace for Peruvian students of engineering. Emilio Guarini, a noted physicist who was a professor at the National School of Arts and Crafts, regularly took his students to visit both plants. Guarini considered Yanacoto to be one "of the most modern and best conceived" plants that he had seen and raved that LL&P would constitute the most powerful trust in South America. "It will only produce positive benefits; will cheapen transport costs, will improve lighting and motor force services; will reduce the price of labor and will favor industry. In few words, the electric industry will be the main factor of Peru's development."[28]

Despite Guarini's prophecy, the electric industry faced many obstacles in the coming years, particularly as electric infrastructure became the object of political and labor disputes. Both plants were in private hands, but groups hoping to influence—or capture—the Peruvian state quickly set their eyes upon them. After Durand's failed plot in 1908, an outright attempt to seize the hydroelectric plants would not take place for another two decades, but the use of electric infrastructure to gain political leverage became commonplace for Lima's burgeoning anarchist labor movements of the early twentieth century. Labor disputes were common in the electric train industry, and strikes took place even before the formation of the company. In December 1906—two years before Durand's ill-fated expedition to Chosica—tramway workers went on strike, resulting in the intervention of then president José Pardo y Barreda. A mild reformist, Pardo, who was not unsympathetic to improving labor conditions, ordered a raise in the workers' wages.[29]

In 1912, tramway workers again went on strike, but this time electric workers joined them. The result was a partial blackout that lasted for a week, causing alarm among Lima's residents.[30] As the strike went on, with the Prados unwilling to negotiate, the "agitating committee" began to pass out leaflets stating that the "insolence and arrogance represented by the exploiters of Lima Light and Power have not only defied the feelings of sensible people, but the solidary feelings of the working classes. . . . we must demonstrate that the cause that is so stubbornly defended by the drivers, motorists and electricians is the cause of all."[31] Light to the city was restored only on the eve of the inauguration of President Guillermo Billinghurst, whose populist tendencies had earned the support of Lima's popular classes.

While romantic attempts at revolution and labor disputes certainly affected the company, the most serious difficulty that LL&P faced was the generation of electricity itself. Between 1914 and 1920 LL&P expanded the production of both Chosica and Yanacoto by replacing old turbines and generators. But the company struggled to keep up with the growth of the city, especially during the First World War. During a visit to Peru, the president of the Pan American Union, Leo Stanton Rowe, noted that while the war had cut off the country from Europe, causing the decline of both Peruvian exports and imports, it had not affected "the cotton spinning factories, which produce the coarser grades of fabric for local consumption."[32] These were precisely the factories that LL&P provided energy to. More importantly, the company, which had originally been created to supply power to Lima's factories, also found itself having to provide electricity as a utility to the city's growing population.

UNDER NEW MANAGEMENT

By the beginning of the 1920s, LL&P was struggling to keep pace with the city's growing demand and was facing financial troubles. Furthermore, unlike during the Aristocratic Republic, the company found itself politically out of favor. Augusto Leguía, who had been president from 1908 to 1912, returned to Peru in 1919 and assumed dictatorial powers, inaugurating the period known as the Oncenio. The Prados, representatives of the oligarchic Civilista Party , were now at odds with the new president, for Leguía—a former member of the party himself—had returned as a staunch anti-*civilista*. Mariano Prado, no longer in good health, resigned and Italian interests took over the company. Under such circumstances, Gino Salocchi, the new head of the Banco Italiano, became a member of the board of directors and sounded out the possibility of hiring electricity magnate Juan Carosio to manage the company.

Born in the Piedmontese province of Arona in 1876, after Italian unification, Carosio had studied at the Swiss Federal Institute of Technology in

Zurich, and after spending a short time in Germany, set out for the "New World." His first forays were in the Argentine electric industry, where he founded the Compañía Ítalo-Argentina de Electricidad. After Carosio's success in Argentina, he made inroads in the Paraguayan electric industry in 1911. As part of Asunción Tramways Light and Power, he was granted a concession to construct and operate electric power stations, for lighting and other industrial uses, in the Paraguayan capital and neighboring towns.[33] As Carosio gained the reputation of the "King of Electricity in South America," Salocchi and Leguía extended him an invitation to guide LL&P's expansion program in 1921.[34] Carosio's arrival demonstrates how the growth of the Peruvian electric industry was aided by regional networks that had been developed in the Southern Cone in the beginning of the twentieth century.

Arriving in Peru, Carosio had two goals. First, to modernize the company's installations and, second, to solve its troubled finances. To acquire capital, he formed the Latina Lux company, based in Milan, bringing together Italian and Swiss capital. He then cemented a partnership with the Motor Columbus holding and the Brown, Boveri & Company, both based in the Swiss town of Baden, who would oversee Carosio's activities in Argentina, Paraguay, and Peru.[35] Swiss investment proved to be so considerable that another company, Compañía Sudamericana de Electricidad (SUDELECTRA), was established in Zurich. SUDELECTRA was headed by Walter E. Boveri, the son of Switzerland's most noted engineer. From the midtwenties onward, he also held a seat on LL&P's board of directors.[36]

Swiss capital was dominant, but Italians were in charge. The new management of the company included Pietro Vaccari, a graduate of the Polytechnic University of Milan, who had collaborated with Carosio in the Argentine electric industry. In 1926 the economist and former civil servant at the Italian Ministry of Public Works Gino Bianchini arrived to speed up the process of reorganization.[37] Self-declared fascists, Bianchini and Vaccari proved to be controversial figures, and although their political sympathies did not become a problem in the early days of the Oncenio, they would be fodder in Peru's political battles in the coming years.

Under Carosio's direction and with the exodus of the Prado family, Leguía doted on the company. LL&P remained an anomaly during the Oncenio: a company controlled by European capital at a time when American investment flooded the country. The company responded by lighting up the capital during the country's centennial celebrations, used by Leguía to showcase Lima as a paradigm of urban modernity. In 1922 Leguía passed a controversial new law that extended the company's monopoly for twenty years. The law exempted the company from import duties for the next five years and exonerated it from having to pay any new taxes for the period of the concession. In return the company was expected to modernize its plants and its grid and expand tramway services.[38] The company duly complied, and between 1922 and 1927 LL&P established an underground system of 10,000-volt cables, upgraded Santa Rosa's distribution control board, and completely modernized the Yanacoto and Chosica plants.[39]

Despite the company fulfilling its obligations, its monopoly status was questioned by critics of Leguía's regime. An early detractor was engineer Víctor Arana, who stated that LL&P's actions should be regulated. In many ways, Arana was not mistaken, as the company's urban infrastructure reflected its laissez-faire origins. Tariffs were unpredictable since there was no distinction between industrial and domestic use of electricity and, more importantly, because no meters were used. Voltage regulators were rare, resulting in either poor lighting or power surges leading to deadly accidents, and the company was less than helpful when attempts were made to inspect their installations. Finally, when Arana promoted the construction of domestic appliances, the company rejected selling them, favoring Italian suppliers.[40]

The company was almost immune to such criticism during the oncenio, but the coming of the Great Depression and Leguía's subsequent fall in 1930 placed it in a precarious position. Given the company's close association with the previous regime, LL&P became the target of attacks by supporters of Luis Miguel Sánchez Cerro, the populist lieutenant colonel who had deposed Leguía. In August 1930 the Tribunal de Sanción Nacional was established, which sought to investigate the corruption behind the public contracts signed during the oncenio. Many companies were denounced in a "moralizing impulse" that sought to soothe an irate public opinion. LL&P was certainly present in the proceedings, but given the ad hoc nature of the court, the company got off lightly: most of those punished were official figures associated with the Ministry of Fomento and Public Works.[41]

The company suffered further blows as it sought to rehabilitate its image in post-Leguía Peru. It had a very public dispute with the mayor of the port city of Callao, Alejandro Roldán, who complained that the company's contractual privileges were unfair and attacked it for raising tramway fares. LL&P fought back, stating that considerable amounts of money had been invested in changing the city's rail network and placing dangerous transmission lines underground, services that were carried out without compensation.[42] Despite this defense, the dispute between LL&P and the city of Callao pushed the government to appoint noted Peruvian engineer Juan Alberto Grieve as inspector of electric services. Grieve also wished to put an end to LL&P's monopoly and for city authorities to take control of the grid, since, according to his estimates, the company was overbilling Lima's consumers.[43] Grieve also accused LL&P of choosing the 220-voltage system over the safer 110-voltage one being used in Europe and the United States, as it was cheaper to install. Grieve's criticisms were characterized as political "demagoguery" by Bianchini, who stated that if the city were to take charge of the company it would go bankrupt within days, as it did not understand the complexities of the industry. He also claimed that the city would go back to the "old ways," in which high-voltage lines killed citizens daily and power was stolen by illegal connections, before the company had invested in modern electric meters, an investment not always welcomed by the city's population as they were rented to consumers.[44]

As the company's disputes intensified, Bianchini became a regular fixture in Lima's various dailies and magazines, and his bouts with Grieve and Roldán were parodied in the Peruvian humor weekly *El Hombre de la Calle* (The Man of the Street), which would publish columns acting as "judge" between the two parties. Although not wholly unsympathetic to the Italian, the weekly judged that "Bianchini should electrocute Mr. Grieve and Mr. Roldán and, afterwards, Mr. Grieve and Mr. Roldán's friends should deliver Bianchini to the beasts at the zoo, who are currently without food."[45] The company's disputes became so notorious, it seems that even Lima's humorists had become exasperated.

Political pressure on the company was eased in March 1931, when Sánchez Cerro was forced into exile, although he vowed to return as a candidate in October of that same year in newly called elections. In the meantime, David Samanez Ocampo, an old politician from the days of the Aristocratic Republic, became interim president. In this delicate political terrain, the company had to deal with the ever more critical problem of labor relations as it sought to apply an austerity program. LL&P argued that the measures were necessary because of the economic difficulties caused by the Great Depression. However, Bianchini and Vaccari had implemented a program that included laying off workers, reducing wages, and cutting work hours even before Leguía's fall.

The company's cutbacks coincided with better organized labor activity under the guidance of the Socialist Party, founded by José Carlos Mariátegui (renamed the Communist Party after his death in 1930), and the anti-imperialist American Popular Revolutionary Alliance (APRA). The two parties quickly became competitors. APRA's ideology—a mix of Marxism and fascism—differed sharply from Mariátegui's, who, despite holding an unorthodox interpretation of Marxism, did not consider APRA to be a truly Marxist party. Despite these differences, LL&P's measures were seen by both groups as a natural process of capitalism, the result of the arrival of Taylorist ideas and the "rationalization of labor" that had characterized the Oncenio. Mariátegui, in his manifesto establishing the General Confederation of Peruvian Workers in 1929, singled out the company, who "in their line of work have lately adopted the system of outsourcing . . . presenting their more qualified or older workers the dilemma of accepting new wages or their immediate firing."[46] Ricardo Martínez de la Torre, a young Marxist writing for Mariátegui's *Amauta* magazine, stated that "LL&P has violently applied the process of rationalization . . . having been careful to fire those whose seniority made them eligible for pensions."[47]

Although the company's measures were primarily directed toward its tramway division—always the least profitable and most problematic of its ventures—it also attempted to fire workers from its hydroelectric plants. In October 1931 the company fired three workers, prompting the "Union of Workers of the Yanacoto, Chosica and Santa Rosa Generating Stations" to mobilize both in the streets and in the press. Appealing directly to the minister of Fomento, the union stated that "the doors of hunger and misery have been violently opened, all of this because we have respectively declined to accept a 25% reduction of our wages."[48] The company claimed that it simply wished to cut hours at the Chosica plant, since the demand for energy had fallen, and that Yanacoto could satisfy Lima's needs.[49] The workers, aware of the precarious position of the interim government, demanded that the state intervene to "avoid producing a social conflict of incalculable proportions which would put an end to the harmony between capital and labor."[50] By now even the State Department paid close attention—even more so after the Schroders had placed 10,000,000 USD worth of SUDELECTRA shares in the market—and conceded that LL&P "has attempted repeatedly during the last year to cut down on operating expenses . . . however, it has been obliged by the circumstances and the threats of strikes and retaliation, as well as Governmental pressure, to suspend its order for more economical operation."[51]

In time, *apristas* would push the communists out of the electric unions. Much as Mariátegui had argued a few years before, the official aprista newspaper, *La Tribuna*, stated that LL&P was simply trying to gradually enforce its long-term plan of rationalizing labor, which would "drag the workers towards violence because misery and hunger *son malas consejeras* (are poor counselors)." The workers—and by extension APRA—ominously stated that they would "not be held accountable of any situation that might take place, given that for some time the union had been precisely fighting for that public calm that the company is trying to jeopardize."[52] When another three workers were fired, and the company established a forty-eight-hour work week, the union took sides with APRA, thanking *La Tribuna* for being the "true and only defender of the national proletariat."[53]

Despite the union's close relationship with the party, the workers did not support APRA's attempts at revolution. Reeling from his electoral loss against Sánchez Cerro after the latter's return from exile, APRA leader Víctor Raúl Haya de la Torre organized a nationwide rebellion, to take place on December 5th. Luis Chanduví Torres, a twenty-two-year-old deputy sergeant and staunch aprista, was directly informed that an uprising would take place, but that Lima would have to fall first. Taking a page from the Bolsheviks' strategy of targeting infrastructure during the Russian Revolution, the uprising was to begin by capturing the Yanacoto hydroelectric plant with the help of the Chosica authorities, led by APRA militant Pedro Bedoya y Villacorta. The uprising was unsuccessful, as apristas "did not take into account the efficiency of the Santa Rosa plant, but only that of Yanacoto."[54] A nervous Chanduví asked Haya de la Torre what had happened in Lima. "Nothing," answered a disillusioned Haya, "it must have been a complete failure . . . The sign to initiate the rebellion was a total blackout, but the lights did not go out." A frustrated Chanduví answered, "People will lose faith in us, next time they will not turn up."[55] Indeed, after the ill-fated Trujillo Rebellion of 1932, apristas continued to cause problems for the government, but they would not launch a direct assault on the government until 1948, with equally disappointing results. But the party's actions symbolized the Peruvian state's newly acquired vulnerability. Even if the state did not build or operate the infrastructure itself, its very stability and legitimacy depended on keeping hydroelectric installations out of the hands of revolutionaries.

Labor relations, already critical, worsened during Sánchez Cerro's period as constitutional president (1931–1933). In June 1932 the company once again tried to enforce a salary cut. This time the company's finances were in crisis, as omnibus and jitney services had affected tramway revenues. The workers threatened to strike if the pay cut was not withdrawn and declared Gino Bianchini a persona non grata. Before another strike went into effect, Sánchez

Cerro held a meeting with the minister of Fomento, Bianchini, and representatives of the electric union. The meeting was possible because the Ministry of Fomento and Public Works was also responsible for labor relations, another arbitrary responsibility that had been adjudicated to the institution. Once more, despite not owning any of the hydroelectric plants, the Peruvian state had to deal with the political consequences brought about by the arrival of hydroelectricity. In its role as an arbiter, the state convinced all parties to agree to an independent commission led by Fernando Gazzani—an old liberal politician who had been close to Augusto Durand and had impeccable anti-Leguía credentials—that would study LL&P's request to raise tariffs and cut wages.

In October another crisis erupted when LL&P fired Guillermo Vargas Irrazabal, leader of the electric workers' union. According to Martínez de la Torre—a by no means impartial observer—Gino Bianchini *wanted* to incite a strike and attempted to utilize Vargas Irrazabal's firing to do so. The "fascist" Bianchini, seeing that the workers were going to present demands that affected the whole of the proletariat, outmaneuvered them by firing Vargas, so that workers would focus only on his reinstatement. By doing so, it was Bianchini's aim to expunge from the union the "most advanced elements of the proletariat—the communists." Indeed, Martínez de la Torre considered Bianchini's support for the rationalization of labor and his fascist tendencies to be one and the same. "Bianchini is a fascist; his methods, his intentions, are fascist. Fascist politics act in such a way that the worker will be subjected to the will of bourgeoise."[56] In the Great Depression, the followers of the late Mariátegui viewed fascism as the ultimate expression of capitalism, and LL&P its most notorious representative in Peru. The real origin of the dispute, however, seemed to lie elsewhere. Martínez mentioned that "aprofascist" elements had been involved. While workers had legitimate grievances against the company, their actions also reflected the competition between communists and apristas within the union.[57]

Sánchez Cerro's presidency did not last long, as he was assassinated by a suspected aprista in April of 1933. His successor, General Óscar Benavides—who would become field marshal in 1940, the only Peruvian to bear this title—brought a degree of political stability, but the company continued to face considerable economic and labor challenges. These economic difficulties had been inherited when Sánchez Cerro had invited the American Edwin Walter Kemmerer as an economic adviser in 1930. The Kemmerer mission sought to carry out substantial monetary and fiscal reforms with the aim of making Peru's economy more attractive to foreign investors after the economic collapse of 1929.[58] One proposal that was adopted was the replacement of the Peruvian pound by the less valuable Peruvian *sol*, causing problems for LL&P, as it had received a considerable injection of Italian and British capital. The company would lament that "monetary depreciation, while it favors local industry and exporters, always negatively affects industrial entities that have to pay in gold loans obtained from abroad in previous years."[59] Even the Marxist Martínez de la Torre agreed with this analysis, although he had little sympathy for the company. "These shareholders cash their dividends in dollars, pounds sterling and Italian liras, while the company receives from their consumers Peruvian soles. With the current exchange rate, they receive less dollars or liras for more soles, hence, despite the company making greater profits in soles, the utilities that the imperialists receive will diminish."[60]

Hoping to raise tariffs and cut wages, Gino Bianchini went on the attack in 1934 by publishing a letter in *El Comercio*, where—despite the newspaper's sympathy for the company—a report claimed that high tariffs had stifled the expansion of electric services in Lima. Bianchini, defending the company from accusations of collusion with the Leguía regime, which had "graciously conceded" LL&P the industrial use of the Andes' waterfalls, stated, "Let us remind you, that these living sources, these natural propulsors only give water, they do not provide machinery or cables, which are imported from abroad, with all the burdens that exchanging the sol entails." The public services that the company offered were the result of two factors: "machinery and human labor." The cost of machinery had been on the rise because of the depreciation of the new Peruvian sol. As for human labor, LL&P had been able to raise wages in 1927 because they had been able to raise tariffs, something that the current government did not allow. Finally, Bianchini, whose fascist tendencies were by now well established, attacked the "philanthropic-humanistic pretext of the working classes and the sacred interests of the people," which did not apply to his company, as over half the houses in Lima and Callao did not use electricity and, in any case, the company had not raised tariffs for working class homes.[61]

In this moment of tension, the Gazzani commission offered its ruling almost two years after it had been established. It determined that the influx of foreign capital, with the arrival of Juan Carosio, had indeed created long-term problems for the company. All foreign loans had been made under the Peruvian pound, and when the currency changed to the sol in 1931, the company found itself on the losing end. Taking this into account, the commission approved LL&P's request to raise tariffs. Furthermore, the state, which had previously received a 50 percent discount on electric services, would now only enjoy 25 percent.[62] The commission's ruling on workers' salaries was far less clear. The company stated the difficulties that it had encountered with its "subaltern" personnel had led to an increase in wages in 1926, which was supported by Leguía in the interest of keeping "social harmony." The company argued that the workers' situation had improved—how this was justified is not known—as the economic position of the company had worsened. The company executives also claimed that they had taken 20 percent pay cuts, while asking the workers to receive only 10 percent less. The commission, however, stayed silent regarding the proposed wage cuts, once again advising the company and workers to reach a mutual agreement.[63]

Given Gazzani's silence, the company announced that wages would be cut by 15 percent. The tramway and electric plant workers' unions gathered in their headquarters downtown "with a crowd that outnumbered those of any previous nights." As other unions joined them, the threat of a general strike became a real possibility. Hoping to avoid a general labor crisis, the Ministry of Fomento once again acted as an arbiter and intervened to help resolve the dispute. Minister Héctor Boza wrote to Pietro Vaccari, asking him to suspend the pay cut. The company, hoping to maintain good relations with the new government, agreed to Boza's request, and Vaccari replied that LL&P agreed with the government's desire to reach an "amicable solution."[64]

The pay cut was withdrawn by LL&P, but the outcome of this last crisis had profound long-term consequences for the company. Most of its labor disputes originated in its tramway division, and, out of solidarity, electric workers would join them. To this end, the American ambassador commented that LL&P's franchise might be modified to retain its light and power activities and turn over its tramways to a separate legal entity.[65] Accepting that its transportation division had been a losing proposition, the company's shareholders decided to get rid of it. The workers, aware that the new company would probably hire them on a lower salary schedule, prepared once more to strike. Hoping to appease labor elements one year after taking power, in 1934 the Benavides government created the public entity Compañía Nacional de Tranvías, which would amalgamate all interurban electric trains. The state was thus no longer an arbiter, but an unwilling owner of LL&P's losing ventures. While LL&P would continue to face difficulties with its electric workers, the demands of the tramway workers were now the state's problem.[66]

The transfer of the tramways would occur gradually in the next few years, but the company immediately changed its English name from Lima Light, Power & Tramways Company to simply Lima Light and Power.

ALPINE-ANDEAN CONNECTIONS

As the Peruvian capital was being rocked by labor disputes, a young Swiss engineer was slowly making his way up the central Andes on muleback. His name was Pablo Boner, and he had been commissioned by Carosio to spot appropriate sites along the Rimac and Santa Eulalia Rivers for future hydroelectric plants. As early as the mid-1920s, the Italian entrepreneur had decided to embark on an ambitious program of expansion as LL&P's existing infrastructure—already functioning at peak capacity—could no longer quench the capital's thirst for energy. Carosio wished not only to meet current demand but to develop an integral plan through 1975 for "the Lima of 1,000,000" inhabitants.[67] Boner's expedition culminated in 1933, when he submitted a report that proposed the construction of five power plants in the next five decades. These were the Callahuanca, Moyopampa, Huampaní, Huinco, and Matucana plants, which would be completed in 1938, 1951, 1960, 1965, and 1971 respectively. All the stations were to be gradually constructed—*de manera escalonada* (in a stepped manner)—as the city's population expanded. Verticality would be the key feature of the system, as each plant was to use the very same waters used by a plant placed at a higher altitude, effectively creating a vertical hydraulic chain.

As Carosio gathered the necessary capital and Bianchini and Vaccari clashed with labor, Boner oversaw the technical aspects of the company's expansion. Born in the small town of Maienfeld in the mountainous Canton of Grisons in 1889, Boner had studied engineering at the Swiss Federal Institute of Technology. After the First World War, he found work with the French government, who posted him to survey water sources in Indochina. He decided to cross the Atlantic and arrive in Asia via the Panama Canal. During his journey, in 1926, he stopped in Lima to visit Ernest Ochsner, another Swiss engineer, who was working at the Yanacoto station. Ochsner, who was to return to Europe shortly, asked Boner to stay. Quickly developing a fascination with the Andes, Boner did not require much convincing and accepted Ochsner's offer.[68] Boner found many similarities between his hometown of Grisons and the Peruvian Andes, something that would also be recognized by his fellow *Bündners*, as "there is something distinct of the Rhaetian region that also overflows abroad and to distant continents." In time, the Peruvian feats accomplished by the Swiss engineer from Graubünden would be "in no way behind the monumental expansion of the alpine hydropower of Switzerland," but Boner's vision was still in its early stages.[69]

Much like Carosio, Boner was obsessed with planning. Years later he would submit to the Peruvian Electrotechnical Association one of his few written works, "The Planning of Hydroelectric Works." While there was still a debate regarding the most appropriate source for electric power—water or oil—Boner decisively argued for the expansion of hydraulic generation of electricity. Thermal plants requiring oil, while appropriate for small urban centers, were more costly, "as thermal equipment has a shorter duration than hydraulic turbines."[70] Furthermore, there was a national interest in taking advantage of hydraulic resources, as it would save oil for export, for which there was great demand in international markets, to bring in foreign currency.

Determining the appropriate technology was only the first step toward elaborating an integral plan. Because the "hydraulic economy" had as its purpose "determining the amount of water which will be available for the generation of electric energy," it was necessary to appraise a river's water levels for several years. If low water levels were found, additional studies must be carried out on the collecting basins, to determine the possibility of increasing flows during dry seasons. Only then could a preliminary project be presented, but even this required factoring in the expanding demand for energy for the foreseeable future. "It is not permissible to plan one hydroelectric plant . . . without taking into consideration the integral use of all the falls, as any hydroelectric work must fit into a wider system planned for the future, which takes into account the maximum use of the river for different stepped plants that can be built as demand grows."[71]

Thus, it was Boner's mission to appraise the water levels of the rivers under LL&P's concession zone, the Santa Eulalia and Rimac Rivers. Boner concluded that both waterways showed great potential. "Dropping a hundred or a hundred and twenty kilometers at altitudes of five thousand meters, the rivers of the coast are . . . exceptional, the most advantageous, the most useful in the world for the production of electric energy."[72] The problem, however, was that their flow was extremely irregular. Explaining with a literary quality, he would add that: "The Rimac and Santa Eulalia rise to a destructive level in three months, between December and April, and then decrease for the rest of the year . . . At high tide the waters hit the mountains hard like a hammer and these crumble . . . during the dry season, they become a feeble current, humbly streaming among the cliffs."[73]

The rivers, like Peruvian politics, were chaotic and unpredictable. But unlike politics, the unruly waterways could be regulated and ordered by the company. To find potential sources of water to increase the flow of both rivers, Boner embarked on a six-year expedition to the provinces of Huarochirí and Marcapomacocha, located between the departments of Lima and Junín. Decades later, the official bulletin of LL&P stated that Boner climbed the Andes with no companion "other than his faith," but he was accompanied by Elias Ludeña, with whom he became close friends.[74] Ludeña, a well-to-do muleteer from Huarochirí province, was familiar with the topography of the area. Little is known about Ludeña, but as early as 1908 he is described in a bulletin of the Society of Engineers as the "guardian of the Huarochirí lakes." The historical registry of the Lurigancho and Chosica districts likewise describes him as an influential person.[75] Ludeña thus had vast knowledge of how long it took for the lakes in the region to be replenished after the dry seasons.[76] Without this local figure, Boner would not have been able to traverse the difficult Andean terrain and would have had to begin his measurements from scratch.

Boner did not write any memoirs, so the details of his expedition with Ludeña remain largely unknown. But it seems that Sixto Elias Ludeña—possibly related to the wealthy muleteer—left some clues on the matter when he went on to study engineering at Ohio State University. He was described by the university's newspaper as "a swarthy, black-haired, well-dressed young man" and a "heir to hundreds of acres of Peruvian lands. His father is an exporter of thousands of tons of wool and beef to foreign countries," and he also hailed from the nearby locality of Chosica.[77] Regardless of the possible family connection, during his stint at Ohio State, Sixto Ludeña published the article "Prospecting in the Peruvian Andes," which sheds some light on how to carry out a successful expedition in the Andes, an experience shared by Boner and Elias Ludeña.

According to the young student, once a "favorable locality" had been established—through information furnished by shepherds—the next step was to find "the shortest path to reach the place." This was easier said than done, for sometimes "the path becomes so narrow and is of such a gradient that you cannot ride on horseback, so you have to walk, and if you are of a nervous nature, it will be wiser not to look down the slopes of the mountains, for if you do, you may become dizzy and fall." It was also useful to know the location of sheep farms where one might sleep during the cold nights. Just as important was the "selection of the riding and pack animals." This was to be determined according to the height that one would reach, "horses, mules and burros being the choice for altitudes around 4,000 meters; 'llamas' and *chuscos* (horses born and raised in the Andes) are the choice for higher altitudes."[78] Finally, Ludeña wrote that a prospecting party was to consist of three men: "two of them do the actual prospecting and the third one takes care of the cooking and looks after the llamas and 'chuscos.'" Ludeña concluded by saying: "In regard to the trip itself, I can say that it is full of romance, thrills and beautiful scenery, and if you have taken good care of yourself before starting, that is, if you are physically fit, the 'soroche' (mountain sickness) will not bother you, and when you return you will surely have plenty of things to tell."[79]

Boner did indeed have much to tell upon his return. Land rich in silver was found, and records from the Ministry of Fomento indicate that Boner and Ludeña, with a man named Roger Charles Picard, bought the land in the 1940s.[80] He also became infected with the dreaded *verruga* (Peruvian wart), from which he almost died.[81] More importantly, however, Boner reached a conclusion regarding which of the two rivers—the Rimac or the Santa Eulalia—was the most appropriate for the company's first phase of expansion. Boner concluded that the Santa Eulalia River was more promising for two reasons. First, it had steeper slopes, which assured more violent falls. "There is no other river in the world like the Santa Eulalia, which descends 450 meters in only 10 kilometers and then arrives at the incredible drop of 1,300 meters in 13 kilometers."[82] Second, a greater flow for the Santa Eulalia could be created by exploiting the Huarochirí lake system, located high in the Andes.

Anthropologists have characterized Pablo Boner as a solitary "water hunter," an almost colonial figure whose sole purpose was to exploit Peru's natural resources.[83] But Boner's expedition was a continuation of nineteenth-century state efforts at making its water sources legible. Some of the waters of the Huarochirí lake system had already been dammed for irrigation purposes, under President José Balta, during the previous century, although these works were largely abandoned after the War of the Pacific. It would be Mariano Prado who restored the dams, in 1901, for his fledging hydroelectric activities.[84] Likewise, the Huarochirí water system had been the object of scientific studies by both Peruvian and foreign engineers. In 1906 George Adams, an American engineer commissioned by the government to look into water sources near Lima for irrigation, stated that the "only deposits worth mentioning are the Huarochirí Lakes, which have been dammed to hold 36,816,048 cubic meters destined for the irrigation of the Rimac Valley."[85]

Thinking of irrigation and not taking into consideration the possibilities for hydroelectric development, Adams considered them to be too far from the land that they ought to benefit, as the water loss from the origin to destination was too great. However, for the purposes of hydroelectric development, the plants could be built on the rivers themselves. *El Comercio*, some years later, stated that Boner's expedition had benefited from "important works published by the scientific centers of the capital, mainly the studies carried out by the engineers Stiles, Turner, Jochamowitz and others, some of which date back to the first years dedicated to this problem."[86]

Boner presented a comprehensive plan in 1932. The "biography" of the first plant to be constructed—Callahuanca—was narrated by Hermann Buse, a noted member of the Sociedad Geográfica de Lima, in a commemorative book when the plant was upgraded in the 1960s. Boner had set his sights on the small town of Barbablanca, on the Santa Eulalia River, not far from the central highway in the San Pedro de Casta district. As most of the workers came from the town of Callahuanca, on the opposite side of the river, the project quickly became identified with that name. As Buse described it, it was not the best site to build a hydroelectric plant. *Huaicos*, or landslides, were common during the rainy season between January and April, causing problems for the local population, who became disconnected from the rest of the country when roads were blocked.[87] Despite these challenges, Boner chose the site for two reasons. First, above the plant was a suitable site to store the redirected waters of the Huarochirí system, as it only required a six-kilometer-long tunnel to transport the waters into a small reservoir. It was also the ideal place to receive additional water in the future, once the waters of the Rimac River were diverted to the Santa Eulalia. Second, in keeping with his vision for the integral development of the region, Boner thought that the waters of Callahuanca could eventually be reused by additional plants built at both higher and lower altitudes.

In 1934 the Ministry of Fomento granted LL&P the right to exploit the waters of the Santa Eulalia River, and work began on the plant that same year.[88] The construction of Callahuanca inevitably brought the company into contact with local communities. Many of the inhabitants had arrived after the War of the Pacific, abandoning the town above Callahuanca, named Chauca, as the new location offered more agricultural opportunities and was closer to the capital.[89] The small town quickly felt the company's arrival, receiving electricity before the plant was completed. A fourteen-kilometer-long connection from LL&P's Chosica plant was established to power the equipment needed to build the plant and light the homes of the workers. At the highest peak of construction, 4,000,000 KWh were used. As Buse narrates, the town was completely transformed: "Overnight, the quiet ravine of Santa Eulalia . . . became a human anthill that took control of the river. The incessant noise of the drills soon followed the dry and destructive explosion of the dynamite. 1,200 men, on a ten-kilometer front, worked without truce."[90]

The town was also impacted by other forms of infrastructural development. Thirty-five kilometers of existing roads were improved, and twenty-two bridges were repaired, supplemented by an additional twenty-two kilometers of new roads and nine bridges that were constructed to transport personnel and machinery.[91] Héctor Boza, minister of Fomento, inaugurated the road that linked Lima to the construction site, although the Peruvian state—other than carrying out random inspections—had little to do with the project. The roads built by the company remained at the service of the community after the plant's inauguration, connecting them to Lima via the central highway. Likewise, the town was the only one in its district that enjoyed a clean water supply and a working sewer system. The construction of the hydroelectric station symbolized the arrival not only of electricity but also of other infrastructures.

However, the company would have to deal with other problems in the future. Although everyday life in Callahuanca was revolutionized, the inhabitants had struggled in the past to obtain adequate supplies of water to irrigate their crops. Likewise, while Callahuanca was directly impacted by the company's activities, the rest of the district, including its namesake capital, still lacked access to electricity and other basic necessities.[92] In time, to cultivate good relations with the local inhabitants, as it climbed higher into the Andes, LL&P would develop irrigation projects and offer social services such as schools, although this policy was not institutionalized until some years later.

As for the plant itself, it had all the features that would be used in future projects. First, an intake received the water from the Huarochirí lake system. Built entirely out of granite, the intake could hold up to 10,000 cubic meters of water. It was equipped with modern electric floodgates that regulated water levels. The floodgates were also equipped with new desanders, designed by Boner himself, to remove the large quantities of rock that the water carried. The intake was connected to a canal made of stone, which in the most vulnerable places was covered with concrete to protect it from incoming *huaicos*. The canal was connected to a six-kilometer-long tunnel through the Andes. This was the most challenging part of the enterprise, as it required twenty-four hours of work daily (divided into three eight-hour shifts) for fourteen

FIG. 6. Callahuanca Power Plant. Hermann Buse, *Callahuanca, 1938–1963* (Lima: EEAA, 1963). Biblioteca Nacional del Perú.

months, using 200,000 kilograms of dynamite and 100,000 electric percussion caps. The waters then arrived at a chamber and were directed toward three steel tubes, each one over one kilometer long, where they would "jump" until they reached the plant, impacting the turbines and generating almost 40,000 KW.[93] The plant was built in such a way that additional turbines could be added, and in the 1950s the generating capacity was expanded to produce over 60,000 KW.

While a plaque commemorating Boner's effort was placed at the entrance of the plant, the station was named after Carosio, who had made possible the influx of Italian and especially Swiss capital that financed its construction. The plant also functioned thanks to Swiss technology provided by Brown, Boveri & Company. In an article titled "Juan Carosio, A Big Modern Hydro-Electric Power Station in Peru," Brown Boveri hailed Callahuanca as "one of the biggest hydroelectric power stations built, so far, in South America" and emphasized that the company had provided all of the electric equipment. Not only did the power station conform to the "very latest conceptions in power plant design," but it also symbolized the end of Lima's dependence on oil. Thanks to the amount of electricity produced by Callahuanca, the Santa Rosa thermal station was kept only "as a standby and peak-load plant."[94] Boner's first step had become a reality.

The inauguration was a momentous event and was attended by President Benavides and his cabinet. The Apostolic Nuncio was present to bless the plant, and everyone in the Italian embassy appears to have attended. Vaccari, Bianchini, and Boner himself went to great lengths to explain to Benavides how the plant worked, as the president was surrounded by *chalanes* (traditional Peruvian horsemen), town authorities, and hundreds of schoolboys. The national press stated that "given the transcendence of electricity in modern life, urban centers require an adequate supply in advantageous conditions for their expansion" but simultaneously highlighted the benefits for the countryside, for the works also favored "the agricultural elements of the valleys close to the capital, for they now have the indispensable amounts of water for the watering of their lands." Finally, the "white coal" benefited "industrial development, brings comfort to human life, disseminating among the whole collective the conquests obtained through modern science."[95]

Vaccari gave a speech in which he thanked Benavides, even though the president had showed little interest in electric development besides dealing with the industry's labor problems. Vaccari diplomatically stated that "the time seems ripe to talk about electric services . . . which might interest a watchful, knowledgeable and modern statesman such as yourself." Though briefly, Vaccari also acknowledged the role played by the Indigenous communities that had provided over 1,200 workers and had maintained "the most cordial of relations" with LL&P and paid homage to those who died or who suffered serious injury in this difficult enterprise. But Vaccari was most thankful to the company's Swiss investors, "who have not collected any dividends for nine years and whose titles have been reduced to 20% of their original value."[96]

Indeed, the Swiss seem to have been particularly patient, especially since, according to Walter E. Boveri, Carosio—who became easily frustrated when the Helvetian investors showed any fiscal restraint—had embarked on the project without informing his partners in Zurich and presented them with a fait accompli.[97] Boveri was nevertheless grateful that Carosio had done so, as the company would have lost their concession had they not been able to satisfy Lima's growing demand for electric power.

The final word, of course, went to Benavides, who appeared to be somewhat humbled during the plant's inauguration. Benavides had decided to prioritize an ambitious road building program as well as creating *restaurantes populares*, hospitals and schools for Lima's small but politically influential working class.[98] Thus, while Benavides characterized Callahuanca as one "of the greatest works of progress carried out in our country," the president also indirectly admitted that the government had not paid enough attention to hydroelectric development by stating that "in Peru, a country of great riches that have barely been exploited, we are seeing a phenomenal growth that more often than not escapes our foresight, as if many of the seeds of its privileged soil suddenly sprouted, already feeling the vital heat of the future."[99] His successor—at least in the domain of electric infrastructure—would not commit the same mistake.

ELECTRIC POLARIZATIONS

Callahuanca proved to be an enormous engineering feat, but the company's technological prowess was overshadowed by the ideological inclinations of its Italian administrators and financiers, especially Gino Salocchi, Gino Bianchini and Pietro Vaccari. Their fascist sympathies had been at the center of the labor battles at the beginning of the decade, and they became the basis of an international campaign spurred by apristas against Benavides after he annulled a presidential election in 1936. Although the Peruvian dictator had attempted to keep both fascists and apristas out of politics, the latter accused the government of being an enclave of the Italian *fascio*. It did not help that Benavides had spent time in Italy during the 1920s, spoke fluent Italian, and, according to American journalist John Thompson Whitaker, the "generalissimo" had become friends with Il Duce. Despite this, Whitaker did not take accusations of fascism against Benavides seriously, characterizing him as an "able and more or less benevolent dictator."[100]

Apristas—who suffered Benavides's benevolence more often than not—were not of Whitaker's opinion, and, while the party at times also flirted with fascism, APRA accused the president of being a fascist stooge. APRA denounced Benavides's association with the "viceroy" Salocchi and claimed that the Banco Italiano influenced all aspects of economic and political life. LL&P—despite having considerable amounts of Swiss and British capital—was likewise accused of being part of this due to the political profile of its Italian administrators. As the world marched toward another global war, electric infrastructure would serve as one of the battlegrounds in which the great powers would vie for the support of the modest Andean nation.

The company's links with fascism remained of little interest to the United States during the pro-American oncenio, but by the 1930s the Department of State showed concern. Salocchi, the main conduit between the Banca Commerciale Italiana and the Banco Italiano in Peru—and, by extension, LL&P and those institutions—was, according to the Americans, the most powerful man in the country. U.S. reports on totalitarian activities stated: "The readiness and ability of Mr. Gino Salocchi . . . to finance any enterprise in which the Italian government evidences the slightest interest is a formidable weapon which is being employed to the utmost."[101] Carlton Beals, in his famous work *The Coming Struggle for Latin America*, also singled out Salocchi as one of the most important financial figures in the country.[102] The only person who could match Salocchi's influence was, of course, Carosio himself. But Carosio had no sympathy for the fascist regime and resisted Mussolini's attempts to bestow honorary distinctions on him.[103] In any case, Carosio avoided becoming embroiled in Peruvian politics.

Salocchi sat on the board of directors, but he held no administrative position in the company. The same could not be said for Bianchini and Vaccari, who in 1939 served as manager and president, respectively. Bianchini's reputation as a financial "wizard" was deemed to be "overrated" by the Americans, who also reported that he yearned to obtain a teaching spot at San Marcos, but that "the university did not care to avail itself of his services." As for Pietro Vaccari, he was characterized as wealthy and "being intensely pro Italian."[104] The Americans were by no means mistaken, as both Bianchini and Vaccari attempted to spread fascist propaganda in Peru. Existing scholarship has highlighted the role played by Bianchini in the creation of the Nucleo di Propaganda, an organization established in 1935 whose purpose was to "combat Anglo-Saxon and Jewish anti-Italian propaganda which had its origins in New York."[105] Furthermore, fascist elements of the Italian colony had the sympathy of *El Comercio*, which, under the direction of Carlos Miró Quesada, openly advocated for the Italian position during the Abyssinian conflict.[106] Not only did Bianchini and Vaccari personally donate money to the Nucleo but LL&P as an institution was its second largest financial contributor.[107]

Apristas used the company's fascist links with great political effect in their fight against the Benavides regime, as demonstrated by a 1938 pamphlet, which was widely circulated in international anti-fascist circles, denouncing the advance of fascist imperialism by the Axis powers. In it, apristas argued that the main aspect of industrial life controlled by Italian fascism was electric infrastructure and singled out LL&P, who "monopolizes the supply of light and motor force in the Peruvian capital and nearby cities," although they conveniently ignored Swiss participation in the company.[108] Due to their anti-imperialist ethos, the party had no doubt that electricity should be controlled by Peruvians, and, indeed, Manuel Seoane, APRA's second in command, considered engineers and industrial development to be critical for the establishment of an "anti-imperialist state" in which foreign capital would be closely controlled to foster national industry, not suffocate it.[109]

The pamphlet expressed apristas' belief that LL&P was engaged in a more nefarious form of imperialism than the United States, its main political target during its first years of existence. It was published shortly before the inauguration of Callahuanca. In it, apristas stressed that the electricity provided by the company was generated "by the falls of the water in the Chosica and Yanacoto plants, and not by oil."[110] The emphasis on hydraulic generation is of great interest. Fascist imperialists were not extracting Peru's natural resources—as the Americans were doing with oil through the International Petroleum Company in the north of the country—but were exploiting Peru's geography to generate wealth. This lent a new angle to Haya de la Torre's views on imperialism. Haya de la Torre famously contradicted Lenin by stating that in the Americas imperialism was the first stage of capitalism, not the last, since foreign powers introduced capitalist dynamics to feudal societies. But by the late 1930s aprista doctrine—despite its own troubled linkages with that ideology—would also view imperialism as the first stage of fascism.

As Benavides's term came to an end, LL&P's political situation became more precarious. Thus far, representatives of fascism and Marxism had been battling it out in Peru's polarized political scene, but newly elected president Manuel Prado Ugarteche would be the representative of liberalism. Prado adopted a pro-Allied stance after his election, breaking off diplomatic relations with the Axis powers in 1942 and openly supporting the United States. According to Prado's first minister of foreign affairs, Alfredo Solf y Muro, Peru shaped its international policy in light of world events, hoping that there would not be the "slightest doubt that the two American nations which are most important to the United States are Peru in the Pacific, and Brazil in the Atlantic."[111] In his quest to move closer to the Allies, Prado considered infrastructural development to be a key factor in helping Peru achieve economic development and better aiding the Allied cause. In this sense, infrastructure, particularly hydroelectricity, played a key role in Peruvian-American relations during the war years.

Shifting his focus away from Lima, Prado sought to industrialize the north of the country through the construction of the Cañón del Pato Hydroelectric Plant on the Santa River—designed by Santiago Antúnez de Mayolo—and a steel mill near the city of Chimbote. He hoped to transform it into the "Pittsburgh of the Andes." Dubbed the Santa Corporation, the project was not borne out of a clear industrial policy, nor was it an effort at economic decentralization. Rather, the project represented an early expression of Prado's personal obsession with electrification. And while the Santa Corporation had a long and tortured history before being shut down in the 1970s, its creation had important international ramifications. Thus far, the electric industry in Peru had been dominated by Italian and Swiss citizens, but now Prado sought the aid of American engineers in his new endeavor.

At first, the Americans were unsure of the merits of the project—they argued that it would produce more steel than Peru needed, not thinking that Peruvians might wish to export it—but they quickly paid attention once the Swede magnate Axel Werner-Gren expressed a desire to invest in it.[112] Americans suddenly realized that supporting the Peruvian state's developmentalist aspirations would ensure support to the Allied cause. After Werner-Gren's visit to Peru, the U.S. State Department actively discouraged the Peruvian government from going into business with the Swede. The Americans informed Prado that Wenner-Gren had been backlisted due to his alleged Nazi sympathies, and a disillusioned Prado stated that "he had been taken by the glitter of the alleged Wenner-Gren millions."[113] Seeing an opening, the Americans supported the recruitment of Barton Jones, an engineer who had been involved with the TVA. The American press celebrated the project, stating that a "scientific revolt in Peru may wreck plans of dictators" and boasted that the "River Santa is capable of producing more and cheaper electric energy than Niagara Falls."[114] If LL&P was a fascist enclave, the Santa Corporation represented the Allied sphere of influence.

The strategy seemed to have worked, and the Peruvian state tried to use American overtures to leverage influence over LL&P. In a letter addressed to the TVA, the government inquired as to the tariff rate charged for the electricity generated by the Boulder and other dams. It was the government's view that in Lima, tariffs were too high for "water which practically has been obtained at a very reduced cost, by building very small dams. For the use of this water, they pay the Peruvian government, yearly, a ridiculous price." While LL&P—and especially Boner—would have shuddered at such a simplistic description of their infrastructure, it was clear that the government was using the company's troubling reputation to its advantage, finishing its letter by stating that the LL&P "was made entirely of Italian financiers."[115]

The United States also sought to secure Peruvian support through the use of propaganda, with the Americans showcasing their scientific achievements through film. The School of Engineers would request several reels from the American Embassy, ranging from the evolution of the oil industry to the linkages between water power and mining.[116] Peruvians, of course, had been aware of these scientific developments since the beginning of the century, but it showed that the Americans were directing their propaganda toward Peru's growing obsession with energy infrastructures. Such films certainly found a receptive audience among the country's engineers, with the Peruvian Technical Association delivering a radio address expressing its support for the United States.[117]

If Americans directed their propaganda efforts at a broader section of society, which included the scientific community, Italians catered to a different type of clientele. The U.S. State Department viewed Italian propaganda as targeting politically influential citizens and thus mainly directed to the upper classes.[118] This would explain why the Italian Cultural Institute, established in 1939 and directed by Salocchi, did not focus on the numerous Italian accomplishments in the field of infrastructure in Peru. Futurism might have been all the rage in fascist Italy, but it did not seem to appeal to Lima's traditional elite, and "typical" Italian themes, such as art and literature, were prioritized over LL&P's cutting-edge technology, present in its sophisticated plants and aesthetically appealing art deco substations. In the end, the institute, which, in the words of Gino Bianchini to José de la Riva-Agüero, had the aim of furthering "intellectual exchange between the two great and progressive nations," closed its doors in 1940, as propaganda against the Axis—and American pressure—proved to be more successful.[119]

Prado's support for the Allied cause was eventually recognized when President Franklin Roosevelt extended him an invitation to the United States in 1942. In an address delivered to Congress, Prado, "slowly and painfully struggling to overcome his natural Spanish accent," stated that Peru was ready to offer "the great possibilities that our territory offers in its rich virgin forests, in its majestic mountain ranges enclosing great mineral and hydroelectric wealth, in its fertile valleys, in its copious rivers, in its mild climate and in the proverbial tranquility of the ocean along its shores."[120] Furthermore, he stated that technical aid would be central to Peru's commitment to the American cause, as "the prompt availability of the equipment and technical direction . . . will allow us to cooperate more intensely . . . to the defense and progress of the continent."[121]

The development of hydroelectricity during the Second World War also pushed Peruvians to rethink their geopolitical position in the region. Having fought a war against Ecuador in 1941, and fearing a Japanese invasion, some considered the Andes a possible refuge in times of war. Military engineer Armando Bueno Ortiz echoed old geographic tropes by stating that the mountain chain created an obstacle to national unity and commercial exchange, but he argued that such challenges could be overcome by its many riches—such as the "powerful and numerous waterfalls that allow for the generation of electricity."[122] It was natural, hence, to consider the Andes to be the "logical" region to be industrialized. The engineer's conclusions were based on the ongoing Soviet experience during the war, as the Russians had contemplated the possibility of retreating from German attacks. The Soviet state could then continue to fight its enemies from the Ural Mountains, which had become "a great industrial region," with iron and steel factories powered by hydroelectric plants. Seemingly surrounded by enemies both close and distant, there was no reason why the Andes could not serve as an industrial refuge in time of conflict as well.

No such existential threats existed, but Prado was willing to give the Allies something more than the economic generosity of the Andes by blacklisting the properties and assets of nationals from the Axis powers. However, because the Italian community had been in Peru for almost a century, becoming embedded in the economy, they suffered far less than German and Japanese nationals. When the idea of blacklisting the Banco Italiano was floated by the Allies, the Peruvian government asked them not to do so as it would create discontent in the country. The case of LL&P was even more complex, for despite being associated with Italian interests, the company also included Swiss, British, and Peruvian capital. Indeed, the memoirs of the board of directors show the heterogeneous nature of LL&P in 1939. While Vaccari and Bianchini continued to hold high positions in the company, the remainder of the board of directors were Swiss nationals with no clear links to the Axis powers. Indeed, while Italians held higher administrative positions, representatives of Swiss capital were becoming more numerous.[123] The highly transnational nature of the company saved it from the extreme measures taken by the Peruvian state, even when its most controversial figures remained in charge. Vaccari eventually stepped down as president but stayed on the board of directors. Bianchini, an even more vilified figure than Vaccari, continued as manager of the company. Even the new president, Hernando de Lavalle, commented observers, had links to the Banco Italiano.

The company also had several Peruvians on the board of directors. Some of them, like Juan Francisco Raffo, had fascist sympathies; others, like Manuel Gallagher, were more difficult to pin down ideologically. To complicate matters further, almost all Peruvian members of the board of directors represented the interests of British shareholders. Gallagher was perhaps the most important member of this group. Although by no means indifferent to fascism—he considered himself an ardent *hispanista* and a supporter of Francisco Franco—he became foreign minister in 1944 and collaborated with the United States in crafting Peru's declaration of war against the remaining Axis powers in 1945. Because of its economic composition, the company did not face the same difficulties as the Banco Italiano, and in any case Italian capital started to disappear during the war years.[124]

Ironically, the person who suffered the consequences of blacklisting was Pablo Boner. Boner never expressed fascist sympathies and, indeed, would later criticize the "arrogance of civilized peoples and of the many mistakes that derive from it."[125] However, as he was unable to prove his Swiss nationality, the Peruvian government considered him to be a German national, which forced him out of the company for the duration of the war. While his family gathered the necessary documentation to prove his nationality, Boner set off for the Amazon, where he founded a business and wrote about the development possibilities of the region.[126]

When the war came to an end, Boner returned to LL&P. There he would find that Swiss capital had taken over the company. Bianchini remained as manager, and Pietro Vaccari continued to sit on the board of directors, but the powerful Walter E. Boveri took the place of Gino Salocchi. These developments did not sit well with the old Italian grandees, such as Bianchini, who in his final years in the company believed that Italians had been pushed out of the company by the Swiss. To add insult to injury, Boveri would receive critical correspondence written in German, correspondence that should have been addressed to Bianchini in Italian. He flew to Buenos Aires to try to obtain Carosio's support on this matter, but the latter, aware of the importance of the Swiss in the company, offered little support.[127]

Despite the ever-increasing economic power of the Swiss, it was clear that the war had negatively affected the company's finances, and when José Luis Bustamante y Rivero was elected president in 1945, exchange controls were enacted that further hampered LL&P's ability to import machinery. As Lima's population continued to grow, LL&P needed to resume Boner's gradual expansion plan. To accumulate new capital, a subsidiary company was created in 1947, named Energía Hidroeléctrica Andina (HIDRANDINA). The board of directors of this new company once again included Juan Carosio and Gino Bianchini, and Swiss figures were also present, with Pablo Boner as one of its executive directors.[128] It would be HIDRANDINA who would finance and build Moyopampa, the first postwar plant, in accordance with Boner's "stepped" approach. Considering Boner's integral vision, the location of the plant was scouted as early as 1936, and the course of the waters that would rush through it was predetermined by the position of the plant at the upper altitudinal level—Callahuanca—and lower level—Yanacoto. Construction began in 1947, and the "planning of the entire installation was entrusted to Motor Columbus," which was once more duly publicized in the Brown Boveri Review.[129] The materials for the transmission line—built by the Peruvian firm Bartolero & Cia—symbolized the resumption of trade with Europe, as they came not only from Switzerland but also from Italy, Sweden, the United States, and even war-torn Germany.[130] Almost rivaling the power produced by Callahuanca, Moyopampa would produce 63,000 KW as the population of the city neared half a million.

Boner's engineering achievements coincided with his compatriots taking full control of the company by the 1960s. Carlos Mariotti and Gaston Wunenburger, the company's new directors, were Swiss nationals. The company also became Americanized in the cultural sense, as it wished to present a friendlier face to the United States as well as to its consumers, especially as the company developed a close relationship with the World Bank. A Peruvianized version of Reddy Kilowatt, a cartoon character created by the Alabama Power Company that gained popularity during the Great Depression, became the company's mascot. It was featured regularly in the company's new bulletin of the same name, *Kilowatito*. (See figure 7.) *Kilowatito* and LL&P became inseparable, as the mascot appeared in all the company's publications and activities, especially the *Feria del Pacífico*, where Peruvian developments in the field of commerce and technology were showcased. The mascot was even adopted by regional electric companies, at times going as far as dressing the beloved character in Andean garb. More critically, unlike in the prewar period, relations between the company's workers and management improved, and the new bulletin had a section on labor relations, which included features on some of its most senior workers and union activities. *Kilowatito* also revealed a close-knit community within the company, as it celebrated sporting tournaments, marriages, and various other social activities.

Despite greater harmony between management and labor, the company continued to face economic challenges. Momentarily unable to construct a new plant at higher altitudes, Boner decided that the only way to increase the generation of existing plants was to divert waters from the Rimac to the Santa Eulalia River. To that end, the "Boner aqueduct" was designed and inaugurated in 1955. Twenty kilometers long—then the longest in South America—the new aqueduct allowed both the Callahuanca and Moyopampa plants to expand their generating capacity, as well as deal with the water shortage problems that dry seasons presented for the company.[131] Fifteen years after the end of the Second World War, LL&P managed to add one more plant to its vertical electrical system. The Huampaní hydroelectric plant, the first of Boner's plants to use the waters of the Rimac River, would produce 30,000 KW, in accordance with the ever-growing energy demand of the city. Huampaní, as *El Comercio* stated, was "in the geographic sense, the last phase of exploiting the scarce but resourceful waters of the Rimac and Santa Eulalia River."[132] With all its infrastructure—including thermal plants—LL&P now generated almost 200,000 KW, representing more than a third of the country's electric power.[133]

As the first chapter of LL&P's history came to an end, it became clear that the company was trying to escape its troubled political past. Yet *Kilowatito* coexisted with the fascist remnants of the company. The Huampaní plant was named after Bianchini, who retired as the company's director after three decades in 1957. Likewise, the canal that fed water to the Callahuanca plant bore the name of Pietro Vaccari. While both Bianchini and Vaccari failed in their attempts to popularize fascism in Peru, LL&P had managed to create a vertical and ordered infrastructural system in the Andes. But the company had reached its vertical limits. The next plant envisioned by Boner—Huinco—would require construction at a higher altitude than the

FIG. 7. First issue of LL&P's bulletin *Kilowatito*. *Kilowatito*, no. 1 (November 1960). Museo de la Electricidad.

company had attempted. LL&P would now require the aid of the Peruvian state who—except for Prado's Santa Corporation—had been largely absent from the electric landscape thus far.

CONCLUSION

Since its inception, LL&P strived to impose order on its economic, geographic, and political surroundings. In its early stages, LL&P was able to break technological and geographical limitations by innovating in the outskirts of the capital as its infrastructure creatively exploited the verticality of the Andes. It also benefited from the flow of capital from abroad, which allowed a small trust to evolve into a large transnational company that could count on financial support to import the necessary machinery to electrify the capital.

But the absorption of capital and scientific know-how from abroad strengthened the political elements of infrastructural development. The actions of LL&P managers—who often were also engineers—show that while

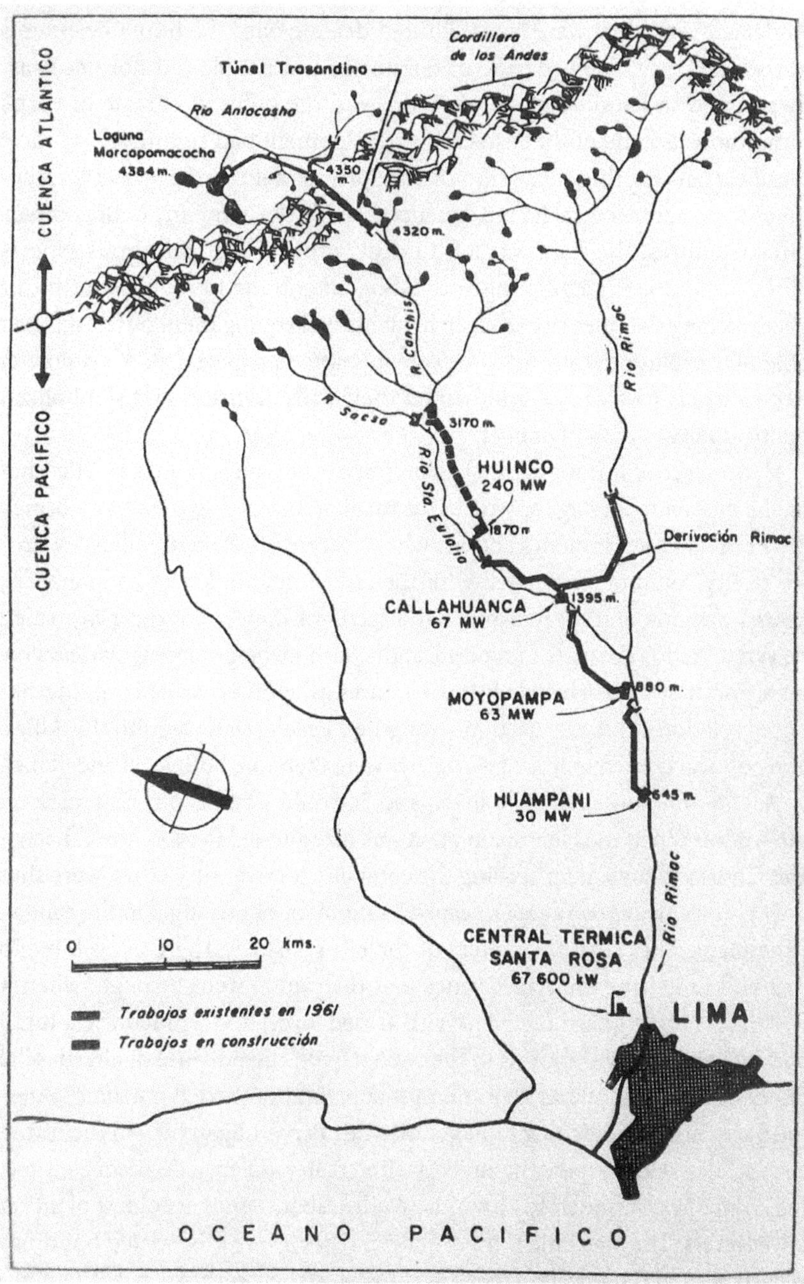

FIG. 8. Vertical depiction of LL&P's hydroelectric chain. *Electrotécnica*, no. 42 (August 1965). Asociación Electrotécnica Peruana.

they thrived at ordering the chaotic nature of the violent and unpredictable Andean rivers, they were less skillful at dealing with the human elements of the company. The ordering of territories and people did not necessarily go hand in hand, as LL&P often misread the political climate in Peru's tumultuous decade of the 1930s. This development had significant political ramifications for the state, although the private nature of electricity sometimes overshadowed such consequences. While the company built complex infrastructures, the state was left to deal with the less glamorous aspects of the sector, be it negotiating with labor, attempting to regulate tariffs, or guaranteeing the integrity of electric plants by keeping them out of the grip of leftist revolutionaries. Lima's residents came to expect a stable supply of electricity, as its absence interrupted their daily routines and symbolized political instability and danger.

Hydroelectric infrastructure had an impact not only on local politics, but on the international stage as well, as the totalitarian leanings of LL&P informed the Peruvian government's actions when crafting its foreign policy. Hydroelectricity could be associated with the fascist infatuation of its managers, APRA's critique of imperialism, or the accomplishments of the TVA, which powered Prado's dreams of modernization. Such ideological imaginaries allow incorporating hydroelectric infrastructures as another variable in international relations as it was used to strengthen Peru's relations with the Allies. The political consequences of electricity were therefore both local and global.

As the company continued to expand, it would leave its physical mark on the Andes. When the Huampaní plant was inaugurated in 1960, the Yanacoto and Chosica plants, after serving the company for over fifty years, were shut down. The closure of Yanacoto caused a moment of nostalgia in the capital. "The plant takes with it chapters of the city's history. There is no Limeño that will forget the uncertain times that the capital went through, when it was necessary to guard the plant with armed force, for if someone captured it, they could bring darkness to the capital, with the purpose of altering the public order."[134] While APRA's attempts at revolution fizzled out after 1948—the party instead choosing to negotiate with Peru's oligarchy—in the future such scenes would repeat themselves with greater violence. In the meantime, the company continued to climb the Andes, abandoning its oldest plants at lower levels. They became infrastructural relics as the dilapidated buildings silently watched the waters that once powered their turbines pass them by.

3

Electrification from Above

"I destroy mountains and save abysses!" shouted Engineer Echecopar, the main character in Enrique Solari Swayne's play *Collacocha*. Inspired by the construction of the Cañón del Pato Hydroelectric Plant in north-central Peru, the 1956 play tells the story of a group of workers in charge of constructing a series of tunnels underneath the Andes in the Ancash Department. At the end of the play, the tunnel connecting the fictional Lake Collacocha with the rest of the system floods, killing over a hundred workers. Little did Solari know that a few years later the workers of Lima Light and Power (LL&P) would attempt the same feat in the Andes of Junín while constructing what was to be Peru's largest hydroelectric facility until the 1970s, the Huinco Power Plant. In the case of Huinco, despite an equally epic flood, the outcome did not end in tragedy. The workers managed to escape, save their equipment, and finish the world's highest tunnel, channeling the waters of the Marcapomacocha lake system into the company's hydroelectric facilities.

The play captured Peru's fascination with infrastructural development and represented "a greater struggle, a reflection of a nation's battle against hostile environmental forces."[1] The story that the riveting play did not tell was the political and economic contradictions behind Peru's hydroelectric frenzy. Experiencing a surge of migration from the Andes in the mid-twentieth century, the city of Lima began to see the rise of *barriadas*, or shantytowns, which increased elite anxieties regarding the social and economic prospects of the capital. The Peruvian government thus embarked on a contradictory process. On the one hand, it encouraged electrification projects in the rest of the country with the hopes of spurring Andean industrialization and stopping the arrival of migrants on the coast. On the other, it needed to supply power to a growing city, a policy that contributed to increasing migratory trends.

While elites were anxious about Lima's future, there was no consensus as to how to change the course of events. This was due to the tension between the two dominant groups in Peruvian society, agrarians and industrialists. The former were powerful landowners whose influence was strongest during the government of Manuel Odría (1948–1956), and who sought to maintain Peru as an agrarian society and were suspicious of any attempt at industrialization. While agrarians did not lose their dominance, industrialists found some political spaces for action during Manuel Prado's second government (1956–1962).[2] However, while Prado favored industrialization, there was no agreement among industrialists as to when and how the state should intervene to do so. With no political consensus on the nature of development, and with no clear political allies among economic elites, the state fell back on infrastructural politics, with the hopes that electricity would alter Peru's economic structures. In the long run this meant that, unlike developmental states, where government and industrialists make alliances, the Peruvian state ended up developing close linkages with infrastructural companies as opposed to industrial groups.[3] And while the development of hydroelectric infrastructures was an engineering success, it remained unclear if it would inevitably lead Peru toward industrialization, despite elites convincing themselves that electricity and development were one and the same.

The Peruvian state's efforts culminated in the elaboration of a National Law of Electric Industry and a National Electrification Plan, the latter crafted by French advisers in 1957, making Peru's hydraulic resources legible to the state for the first time.[4] While the National Electrification Plan signaled Peru's first experience with long-term planning, the only company with enough capital to take advantage of ambitious electrification projects was LL&P, ensuring that hydroelectric plants to power the capital would be given priority. Although national in scale, in many ways the plan reinforced Lima's dominant position. This demonstrated that the benefits of electricity were far from being equally divided between city and countryside, as Andean populations had yet to enjoy substantial access to electric power.[5] It also ensured that the city remained a magnet for Andean migrants, a trend that would become irreversible.

This chapter ends by exploring the transnational linkages, both ideological and material, behind the construction of the Huinco Plant. Singled out as a priority project by the National Electrification Plan, the plant enabled state and company to form an alliance that allowed LL&P to enter into negotiations with the International Bank for Reconstruction and Development (IBRD)—a constituent institution of the World Bank—highlighting once more the importance of large transnational companies in the electrification process.[6] The actions of the IBRD were informed by Modernization Theory—which considered infrastructure a prerequisite for development—reinforcing old elite beliefs that industrialization could be achieved via hydroelectrification alone.[7] Mobilizing considerable human and material resources, as well as scientific expertise, the completion of Huinco was celebrated as a national triumph, reframing the Andes as a place of modernity and Andean peoples as agents of modernization. However, as the media paid greater attention to the company's engineers and administrators, Andean workers, considered "the best in the world," remained largely anonymous.

REGULATING THE ELECTRIC WILDERNESS

In 1948 General Manuel Odría Amoretti carried out a coup d'état against José Luis Bustamante y Rivero. The ousting of Bustamante put an end to Peru's short-lived "democratic spring," which took place between the Second World War and the Cold War, and also to any hopes of economic reform in the foreseeable future.[8] Odría would remain in power for eight years, a period that would become known as the *ochenio*. Shortly after Odría's inauguration as a "constitutional president"—he insisted on holding elections in 1950, which he won after running unopposed—the government issued a document titled *A Program of Action*, outlining his economic and political goals. The document, which was an English translation of a speech given to Congress, criticized the political actions of his immediate predecessor but sought a semblance of continuity with the actions of former presidents. Regarding the fomento sector, Odría vowed to continue the work of Manuel Prado, particularly in hydroelectric development. Hydroelectric plants, according to Odría, would permit "the establishment of new industries such as the manufacture of alkaline and fertilizers, essential for increasing our agricultural production; the production of zinc and, particularly, the exploitation on a large scale of our rich coal resources."[9]

Like previous Peruvian presidents, Odría emphasized the need to industrialize the country. In practice, however, the government's priority was to strengthen Peru's export-oriented model. Even when Odría spoke of expanding the country's electric infrastructure, as the lines in *A Program of Action* seemed to indicate, it was not to promote industrialization, but to better exploit Peru's agricultural and mining resources for export abroad.[10] Two laws were of major importance in this regard. The first was the new Mining Code of 1950, which exempted mining firms from taxes for a period of twenty-five years, allowing the influx of American mining interests, particularly the Marcona Mining Company. The other law was directly connected to the energy sector: the Petroleum Law of 1952, which promoted the search for new oil deposits. This offer was enthusiastically taken by the International Petroleum Company and smaller foreign firms.

Odría, it seemed, was willing to pass new laws regulating any economic activity *except* industry and, by extension, the electricity sector. This can be partly explained by the fact that the National Agrarian Society (SNA) formed the base of his political support in the early years of the Ochenio. The members of this agrarian lobby, according to the American ambassador, "were probably fundamentally opposed to any large-scale industrialization or foreign development of Peruvian natural resources."[11] The SNA had even opposed the passing of the Petroleum Law, for it could potentially create new labor unrest, and oil, even if exported, had an inevitable whiff of industry. Powerful elite agrarian interests, coupled with the American-sponsored mission of economist Julius Klein—which strongly recommended avoiding government intervention on economic matters—temporarily excluded the state from any effective industrial promotion and planning.[12]

This was not a promising start for the electric sector, which languished in the years following the Second World War. It was no longer clear if electric power could be adequately provided as a public utility, let alone as a tool for industrial development. This was particularly true for LL&P, the country's largest electric company. In the preceding decades, the company had constructed a sophisticated hydroelectric system by harnessing the verticality of the Andes near Lima. However, LL&P understood that the altitudinal opportunities for hydroelectric generation near the capital would soon be exhausted. According to the company's expansion plans, it was necessary to climb higher into the Andes, divert more water into its existing facilities, and build another hydroelectric plant at a higher altitude to continue the vertical hydroelectric chain.

But the company could not expand under the existing legal and economic regime. Born in the context of laissez-faire, the lack of regulation had not hampered the growth of the industry. Indeed, such freedom had even been beneficial. By the 1950s the situation had changed. Complex infrastructures were already in place and relations established with consumers, but the growth of the sector still depended on a series of random laws that had been passed in the preceding three decades. At times these laws contradicted each other, leaving electric companies in a legal limbo. Furthermore, the importation of machinery was becoming more expensive as the industry was not exempted from import duties. But as the crisis of the interwar period had demonstrated, one of the main problems was the inability to modify tariffs, as municipal councils oversaw the regulation of prices. It became a vicious circle. Electric companies complained that they could not provide adequate services because tariffs were too low. In turn councils forbade companies from raising tariffs, claiming that the service provided was inefficient.[13]

Without an adequate legal framework, LL&P could not move forward. Reports from the board of directors were somber and pessimistic. The company had been unable to meet its own financial needs or import machinery and materials, leading to the "paralysis of our imports for the efficient maintenance of our installations, forcing us to limit . . . and eventually renounce . . . any extension plans, either to fulfill local demands for electric energy [or] demand from industry in general."[14] Lima's industrialists were alarmed. As early as 1946 the National Society for Industry (SNI) expressed its concerns about the future of LL&P. Given that 80 percent of Peru's industrial activities were in the capital, a shortage of power—which, as the SNI pointed out, was already taking place—would "paralyze our factories."[15] The SNI beseeched the Ministry of Fomento and Public Works to take the company's plight seriously and, by extension, the needs of their industries.

Odría's arrival did bring the company a short respite. Foreign currency controls were abolished, a necessary policy to strengthen Peru's export-oriented model. This allowed the company to import machinery more easily, even if the industrial growth that required electric power was not promoted. Furthermore, LL&P did have friends at high levels of government. The company's president, noted jurist Hernando de Lavalle, would be Odría's ill-fated chosen candidate in the 1956 elections, and Manuel Gallagher, a prominent member of the board of directors, would once more become minister of foreign affairs in 1950. But provisional measures and influential friends were not enough. Indeed, every report of the board of directors from the 1940s through the beginning of the 1950s insisted on the need to properly regulate the industry. In 1950 the company reiterated the "urgent necessity of providing us—and the electric industry in general—with the legal and contractual guarantees that are indispensable . . . for industrial organizations that produce and distribute electricity."[16] It was becoming increasingly clear that the electricity sector could no longer survive in the laissez-faire climate of the Ochenio.

Had the immediate problem been one of promoting industrial development, Odría could have turned a blind eye to the calls of Peru's industrialists, but he could not ignore the energy crisis that the "Greater Lima" area would face in the coming years. By the early 1950s LL&P was generating well over 130,000 KW, but it was expected that by the end of the decade, Lima would require double that amount, as migrants from the Andes flocked to a capital that was nearing one million inhabitants.[17] Fearing the influx of the new residents who erected barriadas, in the outskirts of the city, elites thought these new urban spaces lacked the "most basic principles of morality and hygiene."[18] Ironically, the root cause of this migration was the unwillingness of the same elites, despite their urban anxieties, to carry out a land reform program or promote economic development in the countryside.

Odría could not have addressed the structural causes of this internal migration. This was not only because his political support came from wealthy agrarian groups that opposed economic reform but also because of his disdain for development schemes, especially those in the form of development corporations that had been pursued by Manuel Prado during his first term in office. While unwilling to adapt to the economic reality of the Andes, Odría was aware that support from elites alone would not be enough to remain in power, so he sought to win over the city's new inhabitants. New Andean migrants wanted access to basic services, including connection to the electric grid. Urban modernization, hence, became a cornerstone of Odría's populist policy of winning over the barriadas.[19] However, as more migrants arrived in Lima, electric projects for the capital were given priority over those in the rest of the country. This perpetuated the underdevelopment of the provinces and aggravated the problem of migration.

Thus, the Peruvian state, despite its alliance with agrarian groups, also had to consider the demands of Peru's industrialists—albeit indirectly—through the promotion of a new law that regulated the electric industry. A commission was set up in 1950 that included members of the Ministry of Fomento and Public Works, the Peruvian Association of Electrical Businessmen (AEE), and the Peruvian Electrotechnical Association (AEP). For the next five years, this commission crafted the new legislation. Approved in July of 1955, the law dealt with every detail of regulating the electricity sector. The main objective of the new legislation was to foster both private national and foreign investment, giving the industry the "organic law" that it had demanded for some time.[20] And while the law did not forbid the state from providing electric services, it reaffirmed the private nature of the electric industry, benefiting large companies such as LL&P. As such, the Peruvian law went against existing regional trends, for in countries such as Argentina, Brazil, Chile, and Mexico, public ownership of electric services was becoming the norm.[21]

For the first time since the arrival of this technology, the law officially declared that electricity was a public utility, meaning that the pursuit of electrification was equated with the pursuit of the public good. This allowed electric companies to settle legal disputes that emanated from potential expropriations. Furthermore, it clarified the matter of concessions. For hydroelectric plants, concessions on a given area could last from twenty-five to fifty years, although this applied only to facilities that generated over 100 KW. The law also created an autonomous National Tariff Commission, whose purpose was to reevaluate tariffs on a regular basis, thus taking power away from municipal councils. In an effort to make all electric activity in the country homogeneous, a National Electric Code was drafted. It regulated everything from symbols to definitions and procedures and, more importantly, imposed a uniform current and voltage system for the whole country. Finally, A Supreme Council of Electricity was established to advise the government on the industry's future steps.[22]

After its promulgation, Odría celebrated the law as a personal triumph and, at least in his speeches, seemed to celebrate its industrial implications and its benefits for those living in Lima's surrounding shantytowns. "Peru is no longer at the margins, as it had been before, of recognizing the potential that the electric industry represents, a decisive and necessary factor for urban modernization and industrial development that could not escape my government's attention in its pursuit of progress."[23] The timing of the law was fortunate for Odría. The export boom, which had benefited Peru during his early years in office—much of it driven by exports to the United States during the Korean War—effectively came to an end that year. As the export model faltered, relations between him and the SNA soured. No longer tied to SNA and his followers, Odría wished to transform himself, at least in discourse, into an industrialist.[24]

Odría's political opportunism was not mistaken. The new legislation was celebrated by everyone involved in the electricity sector and associated activities. The American Embassy reported that the law "has been received with great satisfaction in commercial and financial circles," and that it provided "sufficient inducement for long-term capital investment in electrical enterprises . . . and permanent legal guarantees."[25] In their 1955 report LL&P did not necessarily strike a celebratory note, but it expressed relief that the law had been approved. The report informed their shareholders that they could begin claiming their dividends the following year.[26] Some years later, the company would compare the state of the electric industry in the 1950s to that of an orphaned child. Abandoned and unprotected, it had been "scarcely understood by both authorities and citizens, who saw in it only another business, and which grew and progressed thanks to the efforts and courage of a few visionaries."[27]

The law also caught the attention of governments and academics abroad. In 1959 David F. Cavers and James Nelson, professors at Harvard University and Amherst College respectively, published their study *Electric Power Regulation in Latin America*. The study had been commissioned by the IBRD and The United Nations Economic Commission for Latin America (CEPAL). Its goal was to explore the linkages between the electric industry and economic development. By analyzing a handful of countries—Brazil, Chile, Colombia, Costa Rica, and Mexico—both scholars determined that Latin America was characterized by "unsatisfactory" regulations that hampered the development of the sector and, by extension, national economies. Much to their surprise, Cavers and Nelson pointed out that the "leadership in remodeling electric energy laws to adapt them to modern economic developments appears to have been taken by a Latin American country not included in the study: Peru."[28] Indeed, the Peruvian law was studied closely by other governments, not only those mentioned by Cavers and Nelson but also Panama, Nicaragua, El Salvador, and Argentina. It remains a paradox that one of Latin America's least industrialized countries offered the blueprint for regulating such an essential tool for industrial development.

Peruvian academics also celebrated the passing of the law. Economist and geographer Emilio Romero, quoted in a bulletin of the AEE, stated that out of the 57,465 populated centers in Peru, fewer than 300 had access to electric power. Even if the state were to intervene directly to address this problem, it would not have enough resources to solve it, especially considering Peru's other infrastructural needs. The new Law of Electric Industry might, in the short term, benefit companies like LL&P, but it would allow private capital to "decentralize manufacturing and contribute to the prosperity of the Republic." Such decentralization did not manifest itself for some years, however, something that Romero attributed to the "remnants of the preexisting mentality to the electric law, which is making difficult in some places the expansion of electric services."[29]

Those working for the state were also enthusiastic. Fritz Vallenas, representing the Ministry of Fomento and Public Works in his capacity as director of planning at the Directorate of Industry and Electricity, wrote a short essay some years after the fall of Odría divulging the impact of the new legislation. This work is of great interest as it shows how bureaucrats felt about the economic policies enacted during the Ochenio. For Vallenas, the dire situation in the electric sector had been caused not only by the lack of legislation but also because of a "lack of political and economic maturity."[30] Perhaps this was taking aim at Odría's lack of interest in promoting industrial development. The new legislation would now be the cornerstone of Peru's industrial promotion, which was becoming ever more necessary because of the demographic pressures facing the country. The next logical step, Vallenas argued, was state planning. This was already incorporated into the law itself. If the state was to grant concessions, it had to become familiar with its hydraulic resources, Peru's greatest source of energy. Hence, the law also included the formulation of a national electrification plan, which allowed the state to plan—albeit in a limited fashion—the country's economic development.

Even if Odría did not foresee its implications, the law regulating the electric industry was destined, in theory, to promote industrial development. Although this law was usually lumped together with other "organic" laws of the Ochenio—such as the Petroleum and Mining Codes—because it facilitated national and foreign private investment, the very nature of electricity set it apart from other legislation. Unlike oil or minerals, electricity could not be easily exported abroad. Geography and the fact that most plants were far from Peru's frontiers made this possibility unlikely, not to mention the lack of demand in an underdeveloped continent. Therefore, it had to be consumed internally by citizens and industry.[31] Unbeknownst to Odría, he had enacted far-reaching legislation that would allow his successor, Manuel Prado, to push his developmentalist dreams. And while agrarian interests continued to dominate Peruvian society, the industrialists who had been in the shadows for most of the Ochenio could come out into the light.

AN ELECTRIC CONSCIENCE

When Manuel Prado returned to power in 1956 at the head of his own political movement, the Movimiento Democratico Pradista (MDP), he vowed to expand the ambitious development projects that he had championed during his first time in office. These included the Santa Corporation, which had lagged during the Odría years. He would do so by championing an electrification plan that would be national in scale. However, Prado's second term has largely been ignored by scholars, as it did not show concrete results like those in his first time in office, such as Peru's victory over Ecuador in the war of 1941 and his successful pro-Allied foreign policy. Indeed, general histories of Peru portray his second government as one led by a vain aristocrat—who also had to carry the burden of being the son of disgraced president Mariano

Ignacio Prado, known for fleeing the country during the War of the Pacific. It did not help that this second administration was characterized by political turmoil and by economic crisis and subsequent austerity.[32]

Yet, Prado's own speeches, which included clear developmentalist discourse, as well as his actions, set him apart from previous "oligarchic" administrations. Prado's developmentalism had many causes. Unlike agrarian elites, Prado's family had accumulated their capital via financial and industrial activities, and, as shown in the first chapter, his brother Mariano had been one of the founders of the electric industry early in the century. Prado, hence, represented the industrialist faction of the oligarchy, a minority within an already small group. Furthermore, Prado feared Lima's chaotic growth and the expansion of the barriadas—an issue that was gathering weekly headlines and being frequently debated in Congress—which threatened the social structures of the small oligarchic city where he had grown up. Finally, there was one notable difference between Manuel Prado and most of his traditional oligarchic predecessors. While the former were considered "men of letters," Prado had studied engineering and was portrayed as a "scientific" president who could apply his technical expertise to the economic development of Peru.

While much set Prado apart from his predecessors, the president could not escape certain similarities. While embodying the developmentalist aspirations of the Peruvian state, Manuel Prado also represented, in many ways, the ethos of nineteenth-century *civilistas*, as he believed that infrastructure was to be the principal mechanism to spur industrial growth.[33] If, in the nineteenth century, elites believed that railways were the solution to the country's problems, Prado believed that electricity would help Peru overcome the challenges of the twentieth century. With a cohort of engineers that supported his vision, Prado would push for the realization of the National Electrification Plan, which had been approved shortly before his arrival.

Prado attributed Peru's underdevelopment to its lack of industrialization, and the lack of industrialization to a national inability to harness the country's geography to generate cheap energy. This situation had given way to a society characterized by a purely feudal cycle of exploitation "which benefitted the few at the expense of the many," and it was his government's desire that the country could progress toward industrialization, thus "avoiding the colonialist voyage of exporting raw materials so that they return transformed and elaborated as articles to use and consume."[34] This had not yet been achieved because Peru's elites "had not fulfilled the historical task of transforming and integrating its wonderful economic geography" in order to take advantage of the country's cheap energy, which would lead, according to Prado, to a process of "self-industrialization."[35] While his actions in office were certainly not radical, his discourse thundered with language that almost reflected emerging Dependency Theory.

A Gerschenkronian element was also involved, for Peru had imported scientific methods and techniques from abroad yet had failed to use them to escape the trap of economic dependency.[36] "The penetration and absorption of scientific knowledge from abroad in our young nation is fast, not so the mobilization of resources in the field of the economy or the dynamic of a practical will to overcome our stage of underdevelopment."[37] Prado, because of his engineering background, was to inject this "practical will" in the quest for development. Prado's speeches certainly cut deeper than those of any of his predecessors. And while unable to fully translate this discourse into practice and break free from the parameters of what would still be considered a "classical" liberal state in economic terms, Prado did provide the necessary basis for the electrification of Peru. The question remained, as it had in the past, whether infrastructural development would lead to industrialization.

Indeed, the linkages between planning, infrastructure, and industrialization—and when and how the state should intervene to promote these goals—were fleshed out in earnest during these years. As in the case of the Law of Electric Industry, the main forum for this debate was provided by the AEP. As opposed to a lobby group, the AEP was held in high regard because of its academic and technical nature, serving as a space in which industrialists, engineers, and state officials could meet. These included important functionaries of the Directorate of Industry and Electricity—established during the Bustamante years—of the Ministry of Fomento and Public Works, such as Augusto Martinelli, Fritz Vallenas and Fernando Noriega Calmet figures who came closest to embodying a "rational bureaucracy" in Weberian terms. Many of them had joined Fomento during Prado's first government and continued to work in the directorate as late as the 1970s. Likewise, key management figures in LL&P, including Carlos Mariotti and Gaston Wunenburger, were members, as well as notable engineers Pablo Boner and Santiago Antúnez de Mayolo. All of them were active in the AEP's meetings and regularly contributed to its magazine, *Electrotécnica*, which published articles not only on the technical aspects of electricity but also its social, political, and economic ramifications.

By the mid-1950s the AEP had over two hundred members and held regular meetings in its modern locale in downtown Lima. By bringing specialists together, the aim of the AEP was to establish "a direct collaboration between state and private entities linked to the development of national electrification."[38] As we shall later see, this resulted in a peculiar "embedded autonomy," a process that occurs when the state's bureaucracy becomes closely allied with capitalist interests to develop industrial projects. However, the state did not develop an embeddedness with purely industrial entities but rather with infrastructural companies who, nevertheless, sought industrialization, as it would benefit their economic activities.[39]

One year before the promulgation of the Law of Electric Industry, then minister of fomento Fernando Noriega Calmet—a noted engineer and staunch *odriista*—contacted the AEP to solicit their opinion of the new law. The AEP considered it to be a solid legal instrument that would promote private investment but pointed out that private capital would not flow toward remote towns, as they were not attractive markets.[40] As AEP presented its solution to this problem, it quickly became clear that it was more than just a technical body—it also had a political bent.

Some of the most passionate arguments in favor of state intervention were made by Jorge Grieve Madge, a member of one of Peru's most noted "engineer families" and president of the AEP. Much like Prado, Grieve was a product of the nineteenth century. His great-grandfather, Scotsman George Grieve, had arrived in Peru in the 1840s to construct the Tacna–Arica railroad, one of the first on the continent. His grandfather had also been an engineer, who developed cannons for the Peruvian army during the War of the Pacific. Jorge's father, Juan Alberto, followed the family tradition and garnered some notoriety for creating Peru's first national automobile, the "Grieve." When the idea of creating a national automobile industry was presented to then president Leguía, the response from the "diminutive dynamo" was less than encouraging, if not devastating. "We need the products of advanced countries and not experiences with Peruvian products."[41] After the Oncenio, Juan Alberto held the post of Lima's superintendent of energy, clashing with LL&P when he criticized the electric monopoly and called for nationalization of the industry.

While Grieve's father was a well-known figure at the National University of Engineering, and despite his confrontation with LL&P in the 1930s, he remained largely apolitical. This was not the case with his son, who in the 1930s became a member of Alianza Popular Revolucionaria Americana (APRA) and would head the newly established Directorate of Industry and Electricity at Fomento during the Bustamante years.[42] No doubt influenced by his family's victories as well as their defeats, together with his *aprista* background, which had a troubled relation with economic liberalism, Grieve would openly question laissez-faire practices and rejected the idea that the search for private wealth inevitably led to social wealth, as classical liberals argued. Perhaps Grieve summed up Peru's troubled relationship with economic liberalism in one swift phrase: "The paradox of our permanent poverty amidst abundance is proof that economic activity does not regulate itself." Like many other members of the AEP, Grieve celebrated the existing legal regime in the electric sector but believed state action was needed in regions where private investment was absent. Such intervention was necessary to offer light as a public utility and promote economic development.[43]

But state intervention was to be welcomed only if properly planned. This was not a new discovery by Peru's engineers and industrialists. The AEP supported crafting a national electrification plan as early as 1947, and, as such, Peru was on par with other Latin American countries, like Chile and Mexico, that were drafting electrification schemes in the postwar years.[44] The plan stalled during the Ochenio, receiving limited political and financial support—the staff of the planning office was made up of eight people, including clerical personnel.[45] With Prado's return to power, the plan acquired renewed vigor during the first year of his presidency.

Grieve highlighted the importance of planning by quoting World Bank president Eugene Black. If not accompanied by adequate planning, any attempt at development would result in a waste of resources and also lose popular support: "It continues to be true that any development project badly planned or executed does not only consume scarce resources, but also weakens the faith of people in the ability of free governments to produce results that back up their promises." Black would go on to say that this sort of "waste" arose when countries did not pay enough attention to the experts, for they make decisions "on short term political considerations rather than in economic facts."[46]

The future electrification plan, hence, not only affected the electricity sector but highlighted the overall importance of planning. In this regard, eclectic influences were present in the AEP, with different planning models quoted in the various speeches of engineers and industrialists. The Soviet Five-Year Plans were "meticulous and detailed," while the New Deal represented "minimal interference and regulation from the government."[47] The work of Carl Landauer, who in 1947 had written his influential work *Theory of National Economic Planning*, seemed to have a particularly strong impact on the likes of Jorge Grieve, who took to heart Landauer's most basic advice: planning needed a well-defined social goal. For Peruvian industrialists, this social goal was evident. More electricity for Peru meant promoting industrialization and, therefore, improving the living standards of most of its inhabitants.[48] Such aspirations would be the cornerstone for the foundation of the Instituto Nacional de Planificación in 1962, although it would be created not by a democratic government, but by a military regime. As such, the National Electrification Plan was an early example of the planification efforts that would characterize the 1960s.

The crafting of the plan pushed Peru to absorb transnational ideas of planning and put it in close contact with planning experts from Europe, particularly France. Although the plan had advanced slowly during the Odría years, months before the end of his term a deal was struck between the Peruvian government and Électricité de France (EDF) to aid in the realization of the plan. Their arrival was secured with the victory of Manuel Prado, who was obsessed with electricity and, because of his oligarchic background, with all things French. EDF had been active in France and its colonies, Colombia, and Egypt and was currently developing plans for Iran and Turkey, the last two future "developmental" states. The head of the French mission was Marcel Mary, one of France's most celebrated engineers. A graduate of the École Polytechnique, he had overseen the development of the hydroelectric potential of the mountainous *massif central* region in southern France.[49] The remainder of the mission consisted of Daniel Jean Pierre, who had previously worked with Mary; Raymond Wormeringer, who had been active in Algeria; and Maurice Martin, an oil expert.[50]

Before starting their survey, the Frenchmen held regular talks at the AEP to discuss nontechnical problems, as well as deal with "peculiar challenges that are inherent to Peru and that are only known by Peruvians and by those people who have been living (in Peru)."[51] These talks also brought French engineers in close contact with Peruvian technicians. Indeed, the resulting electrification plan was as much Peruvian as it was French, as it included key figures from the Ministry of Fomento and Public Works and members of the AEP, collaborating with dozens of Peruvian engineers, most notably Pablo Boner and Santiago Antúnez de Mayolo.[52] The talks also had the goal of publicizing the mission, for it was understood that the plan would succeed only if public opinion supported it. It was hence necessary, as Abel Labarthe, the then head of the Directorate of Industry and Electricity at Fomento, termed it, to foster a "national electrification conscience," a trait that already characterized the current administration thanks to the "engineer president."[53]

For the AEP, this "electric conscience" meant emphasizing once more that electricity should be not only a public utility but a tool to promote industrial development. Indeed, Abel Labarthe felt that industry had not taken root in Peru because of a lack of electric power, leading the nation to fall into a downward spiral of underdevelopment. "We are in a vicious circle . . . industrialization has not been promoted because there is a lack of electrical energy and vice-versa . . . the electrification of the country has not taken place because of a lack of demand for energy created by industry."[54] To break free from this vicious circle, the government had to offer more electricity than was needed, for industry would eventually demand it. Grieve would even express this in direct political terms: "It is possible to measure the degree of economic evolution of a nation by its consumption of energy. A country that has a limited consumption of energy has a backward economy, and by that fact, is subordinate."[55] Development inevitably became attached to the consumption of energy, underdevelopment to a lack of it. It seems that Prado, and Peruvian elites in general, saw infrastructure as a prerequisite for development.

As these debates took place, the French and Peruvian engineers were hard at work mapping the nation's electric resources. The electrification plan was finalized in 1957, and it outlined the electric potential in Peru for the next twenty years. All projects had to be built by 1975, and a new plan had to be in place before that year. It recommended building fifty-eight new power stations, half of which would be hydraulic. The plan allowed the Peruvian state to make its hydraulic resources "legible," as the nation's untamed and winding rivers were quantified in cubic meters.[56] This legibility led to a new territorial division based on "electric regions." (See figure 9.) In keeping with the state's vision for industrial development, power would flow from nine nuclei. Each of these nine streams of power would flow to its periphery, transforming everything in its path. The nine regions were Lima, the Coastal Central Region, the Central Mining Region, Santa-Trujillo, the Northern Region, the Southern Region, Cuzco, Arequipa, and Iquitos.[57]

The National Electrification Plan may have furthered state power by allowing it to "see" potential hydraulic resources, but the question remained as to who exactly would develop them. Because of its ambitious nature, the plan singled out "priority" hydroelectric projects. This included the expansion of the Cañón del Pato Hydroelectric Plant, to promote Chimbote's steel industry. It also included construction of new plants. Among them were the Pativilca Power Plant, to power the sugar-producing regions north of Lima; a plant in

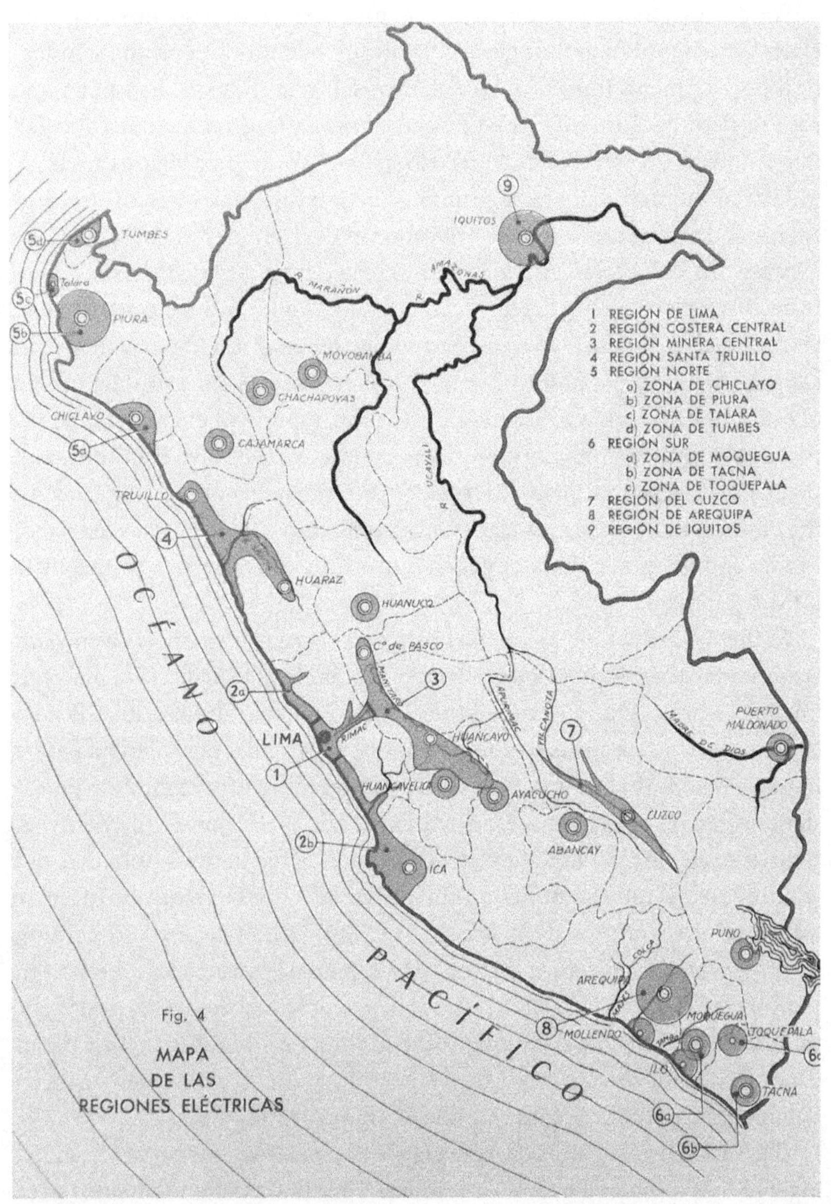

FIG. 9. Peru according to the National Electrification Plan. Électricité de France, *Plan de electrificación nacional* (Paris: J&R Sennac Imprimeurs, 1957). Biblioteca Nacional del Perú.

Machu Picchu, which would help with the reconstruction of Cuzco after the 1950 earthquake; and the Mantaro Hydroelectric Plant, which would promote industrial development in the Central Mining Region.[58] Most urgently for Lima, because of its growing population, it highlighted the importance of building the Huinco Power Plant.[59] The total cost of this last enterprise would be over twenty million dollars.

The AEP supported the creation of a national organism that would build the plants that did not attract private investment. Yet as soon as the plan was published, it highlighted the divisions within Peru's political class over the role to be played by the state. The liberal daily *El Comercio* celebrated the plan, but it also stressed that the state should be cautious in its involvement in the process. Following the parameters of the new electric law, the state should intervene only when projects were too large—or small—for private capital to be involved.[60] *La Prensa* was also highly suspicious when it came to the plan's execution. The daily claimed that the plan gave the state a "far too ambitious" role in the process of development, recommending "thinking long and hard" before embarking on such a risky enterprise, "without previously considering the possibilities of private investment in the field."[61]

The last article was reprinted in the SNI's magazine, *Industria Peruana*, showing the existing divisions within that association. Unlike developmental states in the Far East, such as Japan, South Korea, and Taiwan, Peru's industrialists were not keen on allowing the state to play a central role in the process of industrialization via electrification. There was no *chaebol* in Peru, as these industrial interests were permeated by agricultural ones, resulting in a hybrid class with no clear economic direction. Indeed, even the IBRD would point to this contradiction, stating that one of the main challenges to Peruvian industrialization was "the serious clash of interest between the agricultural and mining capitalist and the industrial capitalist—who, oddly enough, is frequently the same person."[62] The level of economic endogamy between industrialists and agrarians did not allow the state to ally with one group to impose its developmentalist designs on the other. The potential autonomy—the ability of a state to act without being unduly influenced by one class—that could have been gained from this division between Peru's wealthier classes was thus unavailable to the Peruvian state, leaving only infrastructure as an acceptable avenue for state intervention.[63]

The role of the state was also affected by the unwillingness of politicians to support it in its quest for industrialization. This apprehension manifested itself during the crafting of the Law of Industrial Promotion toward the end of the decade. For members of the AEP, the economic development of Peru rested on a triumvirate: The Law of Electric Industry, the National Electrification Plan, and the Law of Industrial Promotion. The moment seemed ripe as well: In 1959 CEPAL—home to Raúl Prebisch, father of Dependency Theory—would publish their first report on the industrial potential of Peru. CEPAL saw many challenges that Peru would face in its quest to promote import substitution industrialization policies, but given the latest government actions, at least the generation of electricity would not be one of them.[64]

The main challenge seemed to be the Law of Industrial Promotion itself. As Thorp and Bertram have shown, the law was unsuccessful, for it granted benefits to all industries on an equal basis, as the law "was deliberately intended to encourage foreign investment, and its unselective nature reflected the prevailing pro-private-enterprise, pro-foreign-capital ethos in Peru."[65] It also showed a clear lack of faith in Peru's own capitalist class. Even Alejandro Belmont Bar, president of the Commission for Industry—when presenting the project to the Chamber of Deputies—stated that "we cannot expect the law to change the way our capitalists invest," although that is precisely what an effective law of industrial promotion ought to have done.[66]

But despite its failure, the Law of Industrial Promotion did spur a debate regarding the bureaucratic makeup of the Peruvian state—an essential part of what Michael Mann would consider a "division of labor within the state's main activities," one of the prerequisites for the development of infrastructural power—in which the relationship between electric infrastructure and industrialization took center stage. Belmont's arguments in favor of the Law of Industrial Promotion had consisted of a poetic metaphor akin to the workings of a hydraulic plant: "To reach our redeeming goals we need a motor faith; make the electrifying and fertile spring of the self-will jump over the rocks of indifference, canalized by knowledge and scientific truth."[67] It was unexpected, therefore, when Belmont and other congressmen proposed that the Directorate of Industry and Electricity of the Ministry of Fomento and Public Works, which had been established in the 1940s, should be split into two. One area would be delegated to industry—specifically, the manufacturing industry—and the other to electrification. This debate quickly dominated the proceedings, showing the extent to which electricity had been associated with industry in the political imaginary, and the extent to which the state's division of labor should be intensified.

Most members of the opposition parties, from both Lima and the provinces, argued against this division. Several congresswomen—elected for the first time to the Chamber, since a 1955 law had given women the right to vote and be elected—proved to be among the most vocal on this point. Alicia Blanco Montesinos, an Independent, argued against the division on the grounds that there was a "close relationship between electricity and industry." This relationship had shown itself during the elaboration of the electrification plan, the goal of which was to not only "supply public and private services to the population, but to develop the country's industrial potential." Another congresswoman, María Eleonora Silva y Silva from Junín—who in the 1960s would fight for the electrification of her department—stated that "electricity without a doubt must be the essential basis of our industry." For many members, electricity and industry had become one and the same.

As for those who did support the division, several congressmen led by Belmont mentioned that Spain, Mexico, Chile, and Argentina, among others, had a section, even a ministry, wholly devoted to industrial development. Yet congressmen such as Juan Sarmiento Espejo of Junín—from Prado's own MDP party—commented that Peru should not create a new ministry simply to fall in line with other nations. Unlike countries in the developed world, and even other Latin American countries, Peru did not have the necessary level of development to have a directorate, let alone a ministry, solely in charge of industry. He was supported in this respect by future reformist Víctor Guevara, who stated that "the bureaucratic examples of other nations are of no use to us."

The final nail in the coffin of the proposed division came from firebrand congressman Héctor Cornejo Chávez, leader of Peru's reformist Christian Democrats. He agreed that there were linkages between industry and other forms of infrastructure but stated that, when it came to industry and electricity, "there is not just a link, there is a causality." He also expressed his argument on technical grounds, for "if the very . . . people in the ministry are against the division, what reason do we have, us who have no technical knowledge on the matter, to argue to the contrary?" But his intervention also showed how even reformists such as Cornejo—who would collaborate with the "revolutionary" government of Juan Velasco a decade later—were reticent about expanding the state apparatus: "Peru is not a rich country; Raimondi's famous phrase, that Peru is a 'beggar sitting on a pile of gold' is entirely untrue . . . the only true part of that metaphor is that we are beggars . . . we cannot afford the luxury of a larger bureaucracy."[68] Víctor Freundt Rosell, a Huancavelica congressman representing Odría's party, was far more succinct, considering the division an "inconvenient bureaucratic development."[69]

It was a point on which agrarians and industrialists agreed: Agrarians did not want more bureaucracy to promote an economic plan that they did not support. Industrialists felt that the link between electricity and industry was essential for the plan to succeed. Furthermore, it showed that the "electric conscience" was not monopolized by *pradistas*, as the link between electricity and industry seemed inescapable for most of Peru's political class. Those against the division won the debate, and the Ministry of Fomento and Public Works retained its current structure until the armed forces took power in 1968.

In the end, the lack of political consensus resulted in the creation not of a ministry but of a sort of agency, dubbed Servicios Eléctricos Nacionales (SEN).[70] This was achieved only when Jorge Grieve—who believed that existing legislation favored only private investment—became minister of fomento in 1960. The purpose of SEN was to rationalize public investment in the sector, particularly in rural areas, and to administer small hydroelectric facilities in places where local authorities were unable to efficiently do so.[71] It should be mentioned, however, that its creation did not truly affect major companies like LL&P or future grand projects, such as the Mantaro Plant, that would have their own autonomous "corporations" during their construction. In the liberal atmosphere of the time, however, the creation of SEN can be regarded as a small victory by those who argued that the state ought to play a bigger role in the electrification process.

Despite all these political maneuverings, Manuel Prado did not lose faith in the National Electrification Plan, stating that it would "allow Peru to enter a stage of industrialization." He also had a clear view of the plan's social purpose, which was to "create work centers in all regions of the country." This not only was an end unto itself but had the larger aim of "preventing human displacement towards the capital and other important centers on the coast, which have given rise to the so-called marginal neighborhoods, causing serious social problems of housing, health and food."[72] Indeed, for Prado, the "emergency" barriadas were a temporary problem, and not a permanent change in Lima's urban landscape. The solution was making sure the "human groups" were tied to "their own places of origin," as Prado claimed that citizens would be "less inclined to abandon the land of their elders . . . if they enjoyed the benefits of an electric installation."[73] Underneath the modern developmentalist discourse, traditional elite fears regarding the future of the capital continued to grow.

Unfortunately for Prado, so did Lima's population, which had almost reached one and a half million inhabitants by the early 1960s.[74] However, as the role of the state in financing various electrification schemes remained uncertain, the only company with enough capital and transnational links to carry out large projects was LL&P. Reinvigorated by both the new legislation and an electrification plan that considered the company's latest expansion—the Huinco Power Plant—a priority project, the LL&P entered negotiations with the IBRD to embark on its most ambitious enterprise yet. Peru's electrification plan might have been "national" in scope and, indeed, it included ambitious projects in the Andes—the Mantaro Hydroelectric project being the most notable example—destined to industrialize the countryside and stop the flow of migrants to the capital. However, unable to escape current demographic trends, Prado would first support a project destined to quench "Greater Lima's" thirst for energy. Peru's first national plan, thus, indirectly reinforced the country's centralist structures, both economic and political.

NEGOTIATING DEVELOPMENT

The possible paths toward industrial growth being pondered by the AEP reflected the evolving ideas about development and its relationship with infrastructure that were being debated in international circles. The idea that infrastructural expansion could lead to economic development was by no means new. Theorists such as Paul Rosenstein-Rodan—proponent of the "Big Push Theory"—stressed the linkages between infrastructure and industrialization as early as the 1940s.[75] This premise was reinforced by Walt Rostow's theory of the "Stages of Economic Growth," a fundamental contribution to what was known as Modernization Theory. Rostow believed that for a society to "take-off" in economic terms, that is, leave traditional economic practices behind and enter the industrial stage, considerable investment in infrastructure—or what he called "social overhead capital"—was needed. These postulates were eagerly taken up by agencies such as the IBRD, the main financier of infrastructural projects. Development was in many ways reduced to the construction of complex infrastructures, thought to naturally lead to an industrial revolution.[76] Despite the novelty of this "new" understanding of development, one may argue that members of the AEP were being informed by both current international trends and past national experience. Peruvians had pushed for railroad construction as a prerequisite for Andean industrialization since the nineteenth century, and for the construction of motor roads in the first half of the twentieth century. Such experiences should have convinced Peruvians that the link between infrastructure and development was anything but clear.

Peru would first interact with the IBRD in 1948, when the institution issued its first report on the country. The IBRD offered a less than optimistic picture: "Geographically, Peru is a basically unhospitable combination of desert, mountain and Amazonian jungle, in which life has never been easy." Such geography had denied Peru "the backbone of a naturally productive agriculture, as exists in the United States, Argentina, or Australia." The greatest irony was that the area had "supported more people in the days of the Incas than it now does, thanks mainly to the efficient, rigid and authoritative control of irrigation."[77] As such, the IBRD echoed a common theme since the beginning of the twentieth century: lamenting that the Peruvian Republic had yet to reach the levels of grandeur that the Incas had achieved.

This pessimistic assessment was offset by the opportunities offered by the very same geography. Peru "is well provided with many natural resources, which permit a diversity of production and of exports not frequently found in Latin American countries."[78] As already stated, one of the problems affecting this potential was the conflict between agrarians and industrialists. One aspect of the clash revolved around the question of wages and manpower. Industrialists sought higher wages to expand the internal market and needed labor for their industries. In turn, agriculturalists complained that "new industries developing in the coastal regions were attracting labor from them, and that new development projects such as road construction, hydroelectric development . . . were similarly stripping the agricultural regions of necessary labor."[79]

The tension between agrarians and industrialists—despite, or because of, the two being closely linked—had impeded Peru's ability to engage with international aid regimes during the ochenio. U.S. Ambassador Harold mentioned as early as 1951 that Peru was not enthusiastic over the possibility of obtaining loans for economic development programs. According to the diplomat, this stemmed from Peru's improved financial position, "which gives them a feeling they can personally finance needed development" and, more importantly, "from a fundamental lack of desire on the part of the agricultural interests for Peru to be industrialized."[80]

The following year the government's attitude seems to have changed, and several applications to the IBRD were made. These loans, however, were not destined to promote industrial development, but rather to "improve its infrastructure and the agrarian sector, both essential to reduce food imports." It is therefore unsurprising that most loan applications fell under the categories of road development, irrigation, agricultural machinery, and expanding mining operations, activities that agrarians found acceptable. There were no applications to expand electrical infrastructure or to develop the manufacturing sector.[81] Even these loans were not approved, as the IBRD was concerned with the amount of debt that the country had already accrued. Peru, who had met its international obligations under the Bretton Woods agreement—it had given its modest quota of 350,000 U.S. dollars—considered the actions of the IBRD to be unjust. Odría complained that Peru's effort "had not been considered by the strong capitalists associated to the entities created in Bretton Woods . . . for we have not yet received the help that we need and to which we have a right, mainly from the World Bank."[82] In time the IBRD would approve these loans, as Odría applied the Klein mission's recommendations and settled Peru's debt.

The nature of loan applications radically changed with the arrival of Manuel Prado. Prado was willing to back ambitious industrial projects financed with international capital, much as he had done during his first term in office. With Latin America's most sophisticated electrical regulation in place and a French-designed electrification plan, hydroelectricity was poised to be the main recipient of international aid. Sanctioned by the government, LL&P began negotiations with the IBRD in late 1958 to finance the Huinco Power Plant, one year after the publication of the National Electrification Plan. Having exhausted the altitudinal force found in the lower levels of the Lima system through its various plants, LL&P's objective was twofold. On the one hand, the company sought to divert new sources of water into the Santa Eulalia River and increase the generating capacity of its existing plants. On the other, it sought to push the vertical frontier by building the Huinco Power Plant and exploit the higher altitudes available in the Andes. This constituted the third stage of expansion envisioned by Pablo Boner—who directed the project himself with technical assistance from Motor Columbus—duplicating the power generated by the company, ensuring the greater Lima area's demand for electricity until the end of the 1960s.

The talks were initiated by Walter E. Boveri, the Swiss partner and most influential figure in LL&P, who was on a first name basis with IBRD president Eugene Black, an American investment banker who had unwillingly become the bank's third president. Boveri did not ask outright for the IBRD to finance what would be the country's largest hydroelectric plant. He first presented a project destined to increase the water capacity of Lima's existing hydroelectric facilities.[83] This entailed the development of the Marcapomacocha water system, a myriad of lakes and ponds found almost 5,000 meters above sea level, in the Junín province of the Andes. The plan was to divert these sources of water through a 20-kilometer tunnel that would join with the Santa Eulalia River. Up to that point, the company had been able to finance the expansion of the system, which had kept it barely ahead of Lima's ever-increasing demand for electricity. Boveri told Black that LL&P no longer had enough capital to finance its completion.[84]

The IBRD was at first unwilling to consider financing the project. LL&P's financial position, although improved, was not enough to gain the bank's confidence. Furthermore, the company's transnational linkages—the very same linkages that gave it considerable cachet in international commercial circles—caused problems for LL&P from the outset. After the Second World War, the company had reinvented itself, leaving behind its association with fascism. While some of the old Italian notables still sat on the board of directors, the company was now managed by Gaston Wunenburger and Carlos Mariotti, both of Swiss extraction. This renewal did not convince the IBRD, who saw LL&P as a Swiss enclave run by the Sociéte Swiss de Electricité Americaine and SUDELECTRA, as well as by Motor Columbus, which in turn were "controlled" by the all-powerful Brown, Boveri & Company. Because of this, when asked about the possibility of a loan, the IBRD asked if Swiss commercial interests would not be a better source of capital. Boveri replied that this would be difficult, for although part of the project had already been financed by selling bonds in Switzerland, the current depreciation of the Peruvian sol made selling more shares unlikely.[85]

Boveri would not take no for an answer. In 1959 he personally wrote to Eugene Black, emphasizing the soundness of the project. More importantly, Boveri knew how to attract Black's attention. He quoted a speech that Black had delivered some months earlier in Mexico City, where he had highlighted the importance of financing utilities. Despite the subject's technical nature, Black stumbled upon a poetic phrase that, according to Boveri, applied all too well to Peru: "The desert was made to bloom, thereby not only reviving the land, but also towns and cities."[86] While the water from the Huinco scheme might indeed be used for irrigation—both government officials and LL&P hoped to profit from this scheme and strengthen bonds with local communities—the towns and cities of Peru's coastal desert, Lima in particular, would bloom because of the electricity of the Andes. Boveri's insistence paid off. The IBRD found itself no longer discussing the Marcapomacocha scheme. According to the bank, it was "now confronted in Peru with a lending program much larger than anticipated. The major addition is, of course, the Huinco hydroelectric project." The total amount was estimated to be between twenty and thirty million dollars. The IBRD committed itself to deciding before the end of 1959 whether it would consider LL&P's request to finance part of the project.[87]

Walter E. Boveri was not the only one involved in the negotiations. LL&P officials were in close contact with the bank, particularly Carlos Mariotti. Their intervention proved to be more useful than that of the Swiss magnate. Mariotti understood that the bank would give priority to projects that required foreign currency. The bank would also consider the fact that the Peruvian economy was based on sound fundamentals and could take on additional external debt. Furthermore, by speaking directly to one of the executive managers of the company, and after sending functionaries to Peru, the IBRD came to the realization that LL&P—one of the last private electric utility companies in Latin America—was very well run. "It is certainly a change to find in Latin America a public utility which has been able to satisfy the demand for power and with a consistently good financial record."[88]

Now that Black and other IBRD officials had been swayed, negotiations moved ahead. Although convinced of the economic and managerial soundness of LL&P and that the project was "feasible and [the] schedule for its development realistic," the IBRD was still unsure as to the willingness of the Peruvian state to back the company.[89] The Ministry of Fomento and Public Works issued a ministerial resolution backing the Huinco project. The resolution was signed by Augusto Martinelli, subdirector of electricity, and supported by then minister of fomento Eduardo Dibós Dammert. The resolution stressed key points that could nudge the bank toward financing the project. First, it highlighted its longevity. LL&P had efficiently provided the greater Lima area with electricity for over half a century. Second, it touched upon the overall legal soundness of the sector in Peru thanks to the new Law of Electric Industry. More importantly, it framed the Huinco project as part of a broader effort by the Peruvian state to modernize the country, as Huinco was declared to be a "priority project," according to the National Electrification Plan. Therefore, Huinco ceased to be an enterprise sought solely by a private company, but rather a project backed by the state to promote development. The document showed synergy between the state and the company.[90]

When the actual amount of the loan came to light, it was the Peruvian government that balked at the sum. In part this was because the government was to guarantee the loan. But, more importantly, Alfonso Rizo Patrón—the new minister of fomento—confided in Roger Chaufournier, the IBRD representative in Peru and eventual director of loans for the Western Hemisphere, that a loan of such magnitude "would affect the country's capacity to undertake other commitments, and he (Rizo Patrón) wanted to check on the relative priority of the project and make sure the company could not raise the money in some other way." President Prado was even more cautious, advising LL&P to reduce its request to ten million dollars. This stemmed from Prado's unfamiliarity with the workings of the bank. Prado was under the impression that the bank had a quota of loans for each country and that any loan would automatically reduce the amount that Peru could borrow. This would mean forgoing other projects such as roads, irrigation, and other hydroelectric plants. While the state was more than willing to back LL&P, it did not want to sacrifice other endeavors in the process. After Chaufournier explained that no such quotas existed, that loans depended on the Peruvian state's capacity to absorb debt, the government withdrew its objections. Indeed, the bank went as far as advising against lending only ten million dollars for it would force the company to look for ten million elsewhere, and the IBRD offered longer-term loans.[91] Development, hence, also meant a learning curve for Peruvian public officials, including its head of state.

Before the loan was approved, the IBRD once more foresaw problems with the company's association with Swiss interests. Every loan required that every stage of the project undergo an international bidding process. However, it was already clear that LL&P would likely push for the use of generators produced by the Swiss company Brown Boveri.[92] The perspective of LL&P was quite different. While the company was open to international bidding for other materials required for the construction of the plant, its position on the use of Brown Boveri generators was that it had technical merit. Most of LL&P's plants already used Swiss machinery, and their technicians were familiar with the equipment. Furthermore, Brown Boveri was establishing a factory on the outskirts of Lima, which meant that spare parts could be easily transported to Huinco's remote location, cutting shipping costs. The company also highlighted a nontechnical aspect. During the 1940s, "back in the years when Lima Light and Power's business was in the red . . . the Brown Boveri Group did not lose faith in our company, and helped it over and over again, technically and financially."[93] Fearful that such a situation might repeat itself in the future, the company wished to remain faithful to its Swiss parent company.

While unhappy with LL&P's decision, the IBRD continued negotiations under the assurance that other equipment would be open to international bidding. As both the requirements of the IBRD and the concerns of the Peruvian state had been satisfied, a twenty-four-million-dollar loan was officially approved in June of 1960, almost six months after negotiations had begun. The official press release stated that the loan was destined to finance two separate but related projects: the Marcapomacocha scheme, to divert the waters from the lakes and rivers to the Santa Eulalia basin, and the construction of a power plant at Huinco, which would generate 240,000 KW, fulfilling Lima's electric needs until the end of the decade. "The new power facilities will increase by 70% the power now available for Lima, the administrative, commercial and industrial center of Peru."[94] Although the Law of Electric Industry and the National Electrification Plan were meant to promote energy expansion and industrial development on a national scale, they proved to be useful tools to secure Lima's political and economic future over that of the rest of the country.

NARRATING DEVELOPMENT

While negotiations between LL&P and the IBRD were anything but simple, the real challenge would be building a hydroelectric plant 4,000 meters above sea level. Construction had already begun in late 1957, roughly two years before negotiations with the IBRD took place, and the financial difficulties faced by LL&P forced it to seek the bank's backing. As the project moved forward, Huinco became the focus of a national frenzy: it was widely publicized by the company and in the press as a great infrastructural achievement that symbolized not only the "conquest" of Peruvian geography but also a giant leap forward in Peru's quest for economic development.

The diversion of the Marcapomacocha system and the construction of the power plant were chronicled in the company's newly inaugurated bulletin, *Kilowatito*, which sought to publicize the company's actions to the Peruvian public. Lima's main dailies, *El Comercio* and *La Prensa*, also celebrated the project, publishing regular updates on its progress. But it would be Hermann Buse de la Guerra who once again offered a meticulous "biography" of the plant in his book *Huinco 240.000 kw.: Historia y geografía de la electricidad en Lima*. Buse's work is an in-depth account of the construction of Huinco and offers a "geography of electricity" particular to Peru. This depicts a struggle of man against nature, incorporating both scientific and mythological elements, as the Andes are presented as a "living" force that man was meant to overcome.

Buse's chronicle begins by focusing on the two rivers that had made possible the construction of the company's complex vertical power system, the

Rimac and Santa Eulalia rivers. By the late 1950s, both rivers—the "hardest working rivers in the world"—were essentially "overworked," as they could no longer increase the energy output of LL&P's existing facilities.[95] To satisfy the demographic and industrial needs of the Greater Lima area, the flow of water had to increase. High in the central Andes, Pablo Boner found that the waters of the Marcapomacocha lake system could be used to augment the flow of the Santa Eulalia River.[96] By setting their sights on Marcapomacocha, however, the engineers of LL&P were pushing not only a vertical frontier but also a political one: the company had left the boundaries of the department of Lima and entered the realms of the Junín department, using its waters to quench the capital's thirst for energy. The foray into the central Andes was also significant since the Marcapomacocha system was connected to several rivers that were tributaries of the Mantaro River, itself a tributary of the Amazon River. In a way, therefore, LL&P was connecting the Atlantic and Pacific oceans. This feat would require the construction of a tunnel 4,400 meters above sea level, then the highest in the world. Just as Peru's central railway built in the nineteenth century had been considered the world's highest, the "trans-Andean tunnel" would be the country's infrastructural claim to fame in the twentieth century. Its construction was depicted as a titanic struggle between man and nature, a glorious effort to "break down the Andean barrier," change the "geographic order," and reach the "virgin" Marcapomacocha system.[97]

Although a desolate place, Marcapomacocha was by no means empty. Before the company's arrival, the area had a small population of one hundred families, most of them devoted to raising sheep, as the terrain in the region was ideal for pasturing. However, as Buse would emphasize in a separate set of articles written for *El Comercio*, the community suffered from erratic access to water. "The god of the altitudes . . . is not always generous" and sometimes "haggled" the waters, or simply denied it to residents for long periods of time.[98] As local communities helped LL&P with the construction of the plant, the company saw it as a matter of "loyalty" to improve local standards of living. As such, the Huinco project was presented as one in which the waters would be used to not only generate electricity but also improve agricultural conditions. In the future, the company made sure that water from its infrastructure was also available to local communities. It was also a key player when the Peruvian and Swiss governments signed an agreement, in 1964, to improve agricultural activities in the region.[99] Apart from water, LL&P also improved local schools and, of course, provided electricity. Such changes affected not only Marcapomacocha's residents but also localities involved in previous projects, such as Callahuanca. This was part of Carlos Mariotti's "progressive" new stand, which would earn the company political points—at least momentarily—with a future "revolutionary" government.[100]

The community would also, as in previous projects, gain road access. As preliminary roads were finished, drilling equipment, generators, fans, pumps, and countless other materials began to arrive in Marcapomacocha. Two camps were erected, Marca and Milloc, on the eastern and western sides of the future trans-Andean tunnel. They were large enough "to house seven hundred people, between engineers, technicians, foremen, administrative personnel, workers and their families, for almost a decade."[101] In early 1958, engineers began drilling, perforating the mountains from both east and west. The project encountered two immediate challenges. The first was the 8,000 cubic meters of rock that had to be cleared and disposed of. It was expected that six locomotives and several wagons would be needed to clear this first obstacle. The second challenge—reminiscent of the Collacocha tragedy—was flooding. Pressing down on the tunnel were the very glaciers and lakes whose waters had to be diverted. As the engineers and workers blasted rock with a generous amount of dynamite, the worst of many floods occurred in February 1962. Buse estimated that 700 liters entered the tunnel per second. In what was considered a "miraculous" escape, no workers lost their lives, and they even managed to salvage their equipment. In these cases, an additional canal had to be built to divert the water caused by the floods. In the quest for water, water itself proved to be the main obstacle.

It was estimated that workers advanced on average five meters a day. Despite the highly technical nature of the work, Buse added an almost mystic quality to the whole enterprise. "Mountains constitute a world apart in geography. What they are, no one knows, not even the wisest scientists. Within them live genies that will challenge men if awaken[ed] from their sleep. Beware those who dare wake them without employing the only two weapons that can assure victory: faith and tenacity!"[102]

This "battle" against geography was being fought with both the tools of Western science and a distinctively Peruvian element, the worker of the sierra. Always referred to as the "anonymous worker of the tunnels," the workers of Huinco were celebrated because of their Andean heritage, as both the company and the press emphasized that only the Andean man could have carried out such an engineering feat. The association between man and the Andes acquired a positive connotation, as the sierra worker—because of his ability to work at high altitudes—became "a true hero of progress." Pablo Boner added: "I will never be eloquent enough to manifest my admiration for his tenacity, his resistance to fatigue and his spirit of collaboration."[103] As for noted doctor Carlos Monge, who decades earlier had stressed the attributes of the "men of the altitudes," he celebrated the racial constitution of the mostly Indigenous workforce in blunt scientific terms: "The man of the Andes has a biological architecture of his own which determines his attitudes."[104] In the context of hydroelectric development, the Social Darwinism that had been used to portray Andean peoples as inferior at the beginning of the century was now used to celebrate them as exceptional beings without whom Peru's modern infrastructural feats would have proved impossible, even if "stronger and more robust" workers from other lands had been employed. What had been needed to recover the ancient Indigenous spirit of thriving in the challenging mountain environment that had been present in pre-Columbian times, it seemed, was the right infrastructure.

Workers again took the main stage in November 1962, when a television version of Solari's *Collacocha* play was broadcasted through Lima's Channel 2. LL&P considered the play "a drama which is intimately linked with the efforts being carried out by our workers in Marcapomacocha." Indeed, the 1962 broadcast included six workers who had been active in the construction of the tunnel. The workers appeared in the play "with their work clothes, helmets, boots, drills, etc., with the aim of giving the drama a sense of realism." LL&P considered the play to be a smash hit, not only because of the actors' magnificent performances but, more importantly, because of the boost in public opinion regarding the work of the company.[105]

In September 1962 the two parties, advancing in opposite directions, finally met. The two leading engineers, who would shake hands to celebrate the connection between the Atlantic and the Pacific, also symbolized the transnational nature of the project. Engineer Juan Hiroshi Endo, the son of Japanese immigrants, was meeting Engineer Saz Fritz of Switzerland.[106] Shortly afterward, the construction of the plant entered its second phase with the creation of the Sheque Reservoir. The reservoir was to be 3,200 meters above sea level and hold 420,000 cubic meters of water. Half a kilometer in length and twenty meters deep, the reservoir would regulate and clean the waters arriving from Marcapomacocha. The final stage of construction would be the titanic task of building a thirteen-kilometer-long pressure tunnel through which the waters would reach the salto de Huinco. At this point the waters "would throw themselves" through tubes that were 1,280 meters tall,

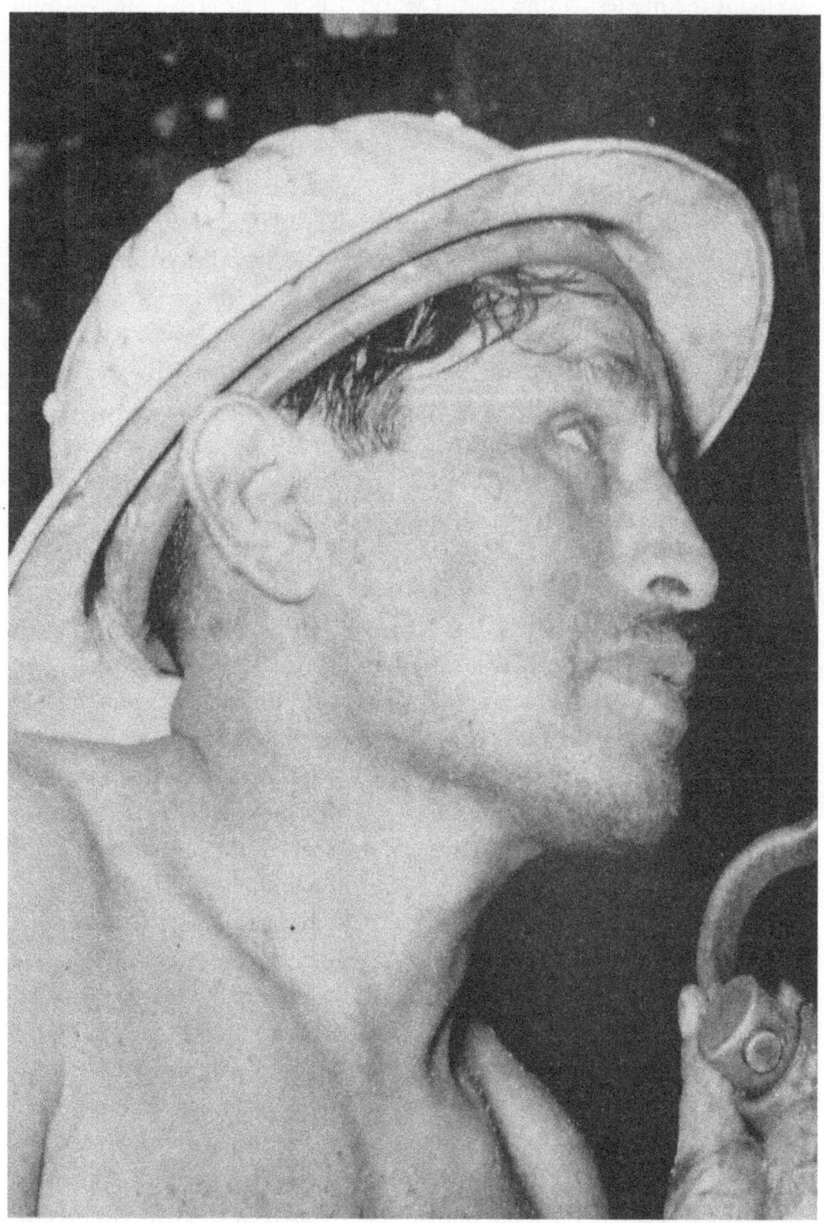

FIG. 10. The worker of the *sierra*. Hermann Buse, *Huinco, 240,000 KW. Historia y geografía de la electricidad en Lima* (Lima: Talleres Gráficos P. L. Villanueva, 1965). Biblioteca Nacional del Perú.

making it the highest "jump" in the world. Upon their arrival, they would pass through a semicircular cavern, moving four turbines, each generating 60,000 KW, quenching the thirst for energy of the Greater Lima Area.[107]

As the plant neared completion, it garnered national attention. Lima's main daily, *El Comercio*, dedicated ten-page spreads to the inauguration, with Huinco featured on the front cover. Each page celebrated the various "landmarks" of the project. The connection between Atlantic and Pacific waters, the highest tunnel in the world, and the Sheque Reservoir were all correctly depicted as great engineering feats. *El Comercio* also celebrated the worker of the sierra, proclaiming that no other worker would have persevered in such conditions. "With water up to his chest day and night, frozen to the bones, coping with irresistible avalanches and swallowing dust from the stone, (the worker) for nearly six years, at 4,400 meters above sea level, wrote a page of value and tenacity, without which Huinco would not be the stupendous reality that it is today."[108] While celebrating the Andean worker was certainly unprecedented in the Peruvian media, it was nevertheless superficial. Unlike LL&P's Swiss engineers and managers, who were mentioned by name and given national honors—Pablo Boner was given the Order of Merit by the Peruvian government, and Carlos Mariotti and Gaston Wunenburger were declared "pioneers" of electricity—the workers of Huinco remained nameless, truly embodying the "anonymous worker of the tunnels." Peruvian elites had judged them to be collectively inferior half a century earlier, and now, as they celebrated them as a group, they were unable to acknowledge the individuality of Andean peoples. It appears that they were seen as the "useful" hands that oligarchs like Francisco García Calderón had unenthusiastically predicted, rather than the creative workers that *indigenistas* had hoped to unleash.

Also, unlike the workers of Huinco, Walter E. Boveri and the IBRD were singled out in the pages of the newspaper, which highlighted the transnational nature of a project that had involved fifteen countries in the free world. Indeed, the transmission line connecting the plant to Lima alone included the participation of countries from four continents. The line itself had been built by SADE, an Argentine electric consortium, in partnership with the Peruvian firm Piazza y Valdez Ingenieros S. A. The towers were constructed by Austrian steel giant Voest, conductors provided by the American company Alcoa, isolators manufactured by Nippon Gaishi Kaisha of Japan and Norden of Denmark, and the armatures imported by the Italian firm Savia. Cranes from the Portuguese firm Mague were used. As expected, the generators were supplied by Brown Boveri.[109]

FIG. 11. Huinco Power Plant: Tubes coming down the mountain. Empresas Eléctricas Asociadas, *60 años de Empresas Eléctricas Asociadas* (Lima: Tip. Santa Rosa, 1966). Instituto Riva-Agüero.

The plant's inauguration took place in 1965. By then the country's political situation had once again changed radically, as Manuel Prado had been ousted by the armed forces in 1962.[110] After a short military junta, new reformist president Fernando Belaúnde Terry was elected and would oversee the culmination of the project. The plant's inauguration would once more cast a light on how Peru's political class conceived the country's identity through infrastructural development. For Belaúnde, Huinco represented "a feat of modern engineering, a victory of science, the victory of a progressive company, and above all, a victory for the Peruvian people." But modern science had not been solely responsible for this feat in the Andes. As we shall see in the next chapter, Belaúnde had based his political campaign on recovering the grandeur of the Inca Empire. Such grandeur, at least in Belaúnde's understanding, had been the Inca ability to "conquer" their geography through great works of infrastructure. If modern Peruvians were to thrive in the Andes and recover the greatness of pre-Columbian times, they had to replicate the great infrastructural works of the past. As such, Belaúnde also emphasized that Huinco had been made possible by "the valuable legacy that we have inherited from our most remote past. That capacity with which ancient Peruvians carried out in our pre-Columbian past—without the technical help available today—admirable works of damming and diverting the waters." Because of that legacy, Belaúnde announced that "the 'Andean Challenge,' as difficult as it is, can be overcome."[111] To the elites, for a brief moment, Peru's "Indians" had become Incas once again.[112]

TABLE 1. INSTALLED CAPACITY, 1950–1965 (KW)

	1950	1955	1960	1965
LL&P	70,000	134,000	197,000	424,000
National	252,000	264,000	582,000	894,000
Total	322,000	398,000	779,000	1,318,000

Source: Empresas Eléctricas Asociadas, *60 años de Empresas Eléctricas Asociadas* (Lima: Emp. Graf. T. Scheuch, 1966).

As for the company, it chose to celebrate Huinco in contemporary terms. Thanks to the inauguration of the plant and the expansion of the generating capacity of other facilities, LL&P had an installed capacity of over 400,000 compared to the almost 900,000 KW of power installed nationally.[113] The company was thus providing its electric quota needed for Peru's "take-off," referring to Rostow's key word for when an economy was entering a period of industrialization. It also presented the company's achievements as beneficial for the Peruvian economy as a whole. Thanks to hydroelectricity, the country would not spend its foreign reserves on oil for thermal plants, which, given Peru's limited supplies, would have been imported. Indeed, Peru's oil would now be available solely for export. For the company, hence, Huinco was essential from a national economic point of view. But even LL&P had to express its Modernization Theory parlance in local terms. As Belaúnde touted the need for Peruvians to "conquer" their own country, LL&P claimed that Huinco, by establishing the conditions for "take-off," was allowing Peruvians

FIG. 12. Peru's electric "take-off." *Electrotécnica*, no. 42 (August 1965). Asociación Electrotécnica Peruana.

to begin to "conquer" development.[114] On the surface, however, it seemed that Lima had "taken off" as the rest of the country remained grounded.

CONCLUSION

With Huinco in operation, *La Prensa* claimed that Lima had become the best illuminated city in Latin America. "The planes that arrive in Lima at night bring large numbers of passengers that are in awe of the city lights."[115] On the ground, however, there was fear for the future of the capital. Long the bastion of elites, Lima was changing rapidly, and the bright lights reflected such convulsions, as some were shining from the precarious homes of newly arrived Andean migrants that were settling in the city's new barriadas. Such were the contradictions of Peru's policy of electrification during these years. The electric industry grew exponentially, as did the power of the state, thanks to the crafting of a national electrification plan, which, for at least some of the elite, showed that Peru had an integral vision of national development that would slow down Andean migration and put an end to such urban anxieties.

But in many ways the plan deepened existing divisions between Lima and the rest of the country, for the very same tools were used to accelerate the construction of hydroelectric projects near the capital, reinforcing internal migratory trends. Elites, fearing that Lima's colonial grandeur was destined to fade, were running out of time as the city's population continued to grow.

Time was not the only problem that Lima's oligarchs faced. Division within Peru's ruling class proved to be equally challenging. Agrarian elites did not want Peru to become an industrial nation, considering that the Peruvian economy, and the country's social order, was of an agrarian nature. Industrialists, on the other hand, saw industrialization as naturally desirable, but the role of the state ought to be limited in this enterprise, in the sense of limiting both its bureaucratic expansion and its direct intervention in the economy. Unable to frame a clear industrialization policy, Manuel Prado and the Peruvian state engaged in infrastructural politics. There was an inevitable feeling of déjà vu, as in the nineteenth century the railroad had also promised Peru an industrial revolution. Unable to learn from past mistakes, Peruvian politicians, industrialists, and engineers equated electricity with industry, and they would convince themselves that the generation of electricity constituted development itself.

The mistaken notion that infrastructure alone could lead to development was strengthened by global ideas such as Modernization Theory. This theory was being propagated by the IBRD, who financed the first project in the National Electrification Plan, the Huinco Power Plant. The state formed an alliance with LL&P, a company whose actions were indirectly associated with industry. But while the inauguration of Huinco meant that Lima's demand for power would be satisfied first, the plant's construction did have an impact on Andean geographical and social imaginaries. Previously seen as an obstacle to economic growth and national unity, Indigenous workers were depicted as the heroes of modernization. For intellectuals and statesmen hydroelectricity appeared to be the "right" kind of infrastructure in which Peru's native population could thrive. It would be one of the few moments in the twentieth century in which these subaltern populations were celebrated because of their racial composition and geographical origins. It remained unclear, however, if such celebration would continue after the electric frenzy was over.

With Huinco momentarily satisfying Lima's demand for energy, elites could now turn their attention to projects in the Andes. In the meantime, the engineering achievement of Huinco was to be celebrated. But such celebrations appeared to go largely unnoticed by the rest of the world, something that was lamented early on by the IBRD, who believed that its projects in South America should garner greater attention. "The strangest thing about the great hydroelectric schemes in South America is that they are so little known. Tourists go elsewhere to search for the last flickering remnants of old civilizations."[116] The country, it seemed, could no more easily escape its pre-Columbian legacy than it could that of the nineteenth century.

4 Electrification from Below

"The true challenge of Peruvian electrification is that of rural electrification."[1] These were the words of engineer Mario Calmet—director of rural electrification at the Ministry of Fomento and Public Works—to his peers during a conference in the 1960s. Calmet lamented that despite the construction of large, sophisticated hydroelectric plants in the Andes, less than 13 percent of people in rural areas received electricity, even though the majority of the country's population lived in the countryside.[2] The challenge, thus, was how to provide power to dispersed rural populations, connecting them to a wider system and to the state and the nation at large.

This chapter examines the impact of rural electrification in the Peruvian central Andes through the history of the Eléctrica Comunal del Centro Ltda. No. 127, which was the result of the developmentalist aspirations of the people of the Mantaro Valley. The successful establishment of an electric cooperative in 1964 was possible due to a synchronicity of national and local discourses of development, an uncommon experience in the central Andean region.[3] At the national level Fernando Belaúnde Terry became president promising to overhaul existing social and economic structures, after decades of oligarchic government, through the recovery of "lost" infrastructural knowledges of pre-Columbian peoples, of which communal cooperation was an integral part. At the local level, elites believed that cooperative development was not only possible but desirable, as it was compatible with the exceptional economic and social nature of the valley. Both national and local figures agreed that cooperatives allowed for the expression of traditional communal practices through what was considered a modern, Western capitalist institution.

The history of the Cooperativa Comunal del Centro also sheds light on the transnational nature of the development of the Peruvian electricity sector, as U.S. representatives found themselves preaching the gospel of rural electrification in Cold War Peru. National and regional discourses were compatible with the aims of American agencies, which acquired newfound importance during the Alliance for Progress, a program created to fight communism in Latin America via economic development. One such agency was the National Rural Electric Cooperative Association (NRECA), which had its origins in the days of the New Deal but came to play a prominent role in the rural electrification of Latin America during these years. The cooperative, thus, allows a rare occasion to analyze the long-term transnational impact of the New Deal outside of the United States and adjacent territories during the Cold War.[4]

Such efforts were enhanced by the actions of the United States Agency for International Development (USAID), whose electrification experts saw in the cooperative a combination of traditional social organization and modern technology. As such, they saw it as a homegrown model of development that could aid Peru in achieving its aspirations to modernize, placing infrastructural development as a key variable around which Cold War battles were conducted.[5] Furthermore, USAID also recognized structures and traditions in Indigenous communities that were compatible with the American cooperative model. In time, the residents of the valley would modify this blueprint through local labor and social practices, creating a hybrid cooperative model that showed that modernization efforts were far from being totalizing endeavors.[6] While contemporary development theories considered "traditional" practices an obstacle to economic modernity, in this case they proved to be an asset for the adaption of the Western cooperative model.[7]

American experts established close alliances with Peruvian officials and, more importantly, the communities of the valley, who saw the state, despite its reformist discourse, as a distant partner, unlike the Americans who interacted with them daily. Bringing to the fore the role of experts in Peru's electrification endeavors can shed light on how Cold War dynamics were experienced in the local context, as experts became embroiled in political disputes, for, unlike earlier electrification projects, which were mostly in isolation, the cooperative required daily interaction with Indigenous communities.[8] Internal battles within the cooperative, tensions between local residents and the central government, and diplomatic disputes with the United States all left their mark on the history of the cooperative, who often had to struggle for financial and technological resources and, most of all, electricity itself.

These tensions came to a climax when Belaúnde was overthrown by a "revolutionary" military government in 1968 amid the fluid political context of Peru's Cold War. Promising a "revolution from above," the new military regime stopped portraying the cooperative as an organization destined to foster communal capitalism. It would eventually depict it as an obsolete institution that perpetuated the structural injustices of the capitalist system. The cooperative's links to the United States would be of no help as relations between the United States and Peru continued to sour. Ironically, a government that promised to empower the peasantry expropriated the cooperative against the community's wishes, leading to the dissolution of Latin America's largest rural electric cooperative, and as they became connected to a larger national grid, the residents of the valley lost the autonomy that they had previously held.[9]

FROM THE ANCIENT AYLLU TO THE MODERN COOPERATIVE

The arrival of the Alliance for Progress in Latin America found fertile ground in 1960s Peru. In 1963 architect Fernando Belaúnde Terry and his reformist Acción Popular (AP) party arrived at the presidential palace. During the electoral campaign, Belaúnde promised to institute radical changes within the context of liberal democracy. Furthermore, the arrival of the new government also raised expectations of an improvement in Peruvian-U.S. relations, which had undergone some tension during the short-lived military junta that preceded Belaúnde's arrival. The new president's own background—having undergone his undergraduate studies at the University of Texas at Austin—seemed to add to these expectations. The Alliance for Progress appeared particularly well fitted for the new reformist government, and Belaúnde seemed to personify the American vision of what a Latin American leader was supposed to be: progressive enough to deter communism, yet not radical enough to promote it.

Belaúnde and his followers had been developing their reformist philosophy for nearly a decade before the arrival of the Alliance for Progress. While Kennedy's ambitious program rested on the postulates of Modernization Theory, presenting Latin America's countries with a linear path toward economic development, Belaúnde stressed the need to find political and economic inspiration in Peru's pre-Columbian past, a decidedly non-modern source. It was in this intellectual milieu that Belaúnde built a political platform around the need to revive ancient practices for the benefit of modern Peru. This was best expressed in his most famous work, *La conquista del Perú por los peruanos*. In it, Belaúnde celebrated the abilities of ancient cultures to develop sophisticated infrastructure in their quest to harness Peru's challenging geography. It proved to be an enormously popular work and would form the intellectual cornerstone of his political movement.

Five centuries had passed since the conquest of the Tahuantinsuyo by the Spanish Empire, and the revival of practices from a world that no longer existed raised the question of how they would be applied in Republican Peru. It was therefore necessary to express the social and economic practices of the Incas through a modern institution. The ancient *ayllu*—the principal form of communal organization in pre-conquest Peru, characterized by self-sufficiency and reciprocity among its members—could find its modern equivalent in the Western cooperative. Indeed, Belaúnde claimed that "Ancient Peru presents us with a regime that is fashionable in the most progressive nations of the Old World. We are speaking of the Cooperative system."[10] The global cooperative movement, which emerged in the mid-nineteenth century when the Rochdale Pioneers established the first consumer cooperative in the outskirts of Manchester, was compatible with pre-Columbian practices that still lingered in the heart of the Andes.

Besides allowing Belaúnde to establish a political philosophy that combined tradition with modernity, the celebration of this ancient cooperative spirit of the Incas also allowed him to present himself as a reformist and not a left-wing revolutionary. "Ancient Peruvians were cooperativists, not communists," wrote Belaúnde emphatically.[11] This point was of no small importance. In the context of Cold War Peru, the debate over whether the extinguished empire was socialist—or, worse still, communist—had considerable implications for contemporary politics, both local and global. Belaúnde was not breaking new ground in this debate. Socialist interpretations of Inca political and economic life and its linkages to cooperativism dated back to before the Cold War, when in 1936 Peruvian intellectual Hildebrando Castro Pozo wrote *Del ayllu al cooperativismo socialista*. A member of the Peruvian Socialist Party, founded by José Carlos Mariátegui, Castro Pozo believed—much like Mariátegui himself—that Peru's pre-Columbian socialist past inevitably pointed toward a communist future, characterized by the combination of cooperatives and modern technology.[12]

The debate also expanded beyond Peru's frontiers and found expression in a work crafted by French economist Louis Baudin, *A Socialist Empire: The Incas of Peru*. A faithful disciple of the emerging Mont Pellerin Society, Baudin took issue with the notion that the Inca ayllu had any link with the modern cooperative and argued that the despotism of the Inca Empire was akin to the socialism of the Soviet Union. In later editions of his book—which had originally been written in the 1920s—Baudin challenged Castro Pozo's "socialist" interpretation of cooperatives: "Incidentally, Pozo's title is incorrect. The theory of co-operatives is one doctrine; socialism is another. A desire to appear socialist is apparently what prompts their juxtaposition."[13] The Empire, extinct for over four hundred years, remained at the center of Peru's political battles, and for proto-neoliberals like Baudin and Ludwig von

Mises—who wrote the preface to the 1961 edition of Baudin's work—at the center of a larger global conflict.

Belaúnde, as a progressive reformist, avoided these extremes. He saw in cooperatives a path to a successful and compassionate capitalism, if not the ultimate expression of capitalist development. Not only did they coexist with individual property, "for the cooperative itself is the owner," but they also had the "advantages of the capitalist system and none of its vices, for it is structured in a way that avoids the dangers of speculation." Finally, Belaúnde, with his usual penchant for grandeur, argued that cooperatives gave "the common man the possibility of organizing with the same efficiency as the great capitalist consortiums."[14]

Belaúnde's ideas found concrete expression in one of his most ambitious programs: Cooperación Popular. The goal of the program was to empower the peasantry to "become aware" of their surrounding environment and natural resources and exploit them in the construction of public works.[15] Peasant communities would undertake the construction of public works at their own initiative, and, in exchange, the state would provide technical assistance and resources to aid them in their endeavors.[16] Cooperación Popular, an essential part of his "popular action" doctrine, once more reflected the amalgamation of the old and the new, for it was a project that "bases its principles on the echoes of ancestral voices, without ignoring the ample technological possibilities of our time."[17] While inevitably linked to cooperative development, the program was not to directly organize and foster the cooperatives. It would assist them once they were established and registered through the newly created National Institute of Cooperatives (INCOOP). The law that created this new institution decreed cooperatives to be a "national necessity . . . an efficient system, contributing to economic development, the strengthening of democracy and the realization of social justice."[18]

By setting up the necessary legal scaffolding and institutional framework, Belaúnde sought to address the challenges that Peruvian cooperatives had faced in the past. The first phase of "modern" cooperative development took place during the 1940s and 1950s and was dominated by credit unions, the most famous of these being established by Fr. Daniel McClellan of the U.S.-based Maryknollers in Puno. However, despite the isolated success of this "neocapitalist" experiment, it failed to become a national movement.[19] Years later, a report of the executive office of the USAID explained that the cooperative movement had not been successful during these years because "it appeared to lack a sense of direction."[20] The failures could be mainly attributed to the lack of a cooperative law. A law passed by President Manuel Prado in 1943 did not go far enough, as it merely recognized the judicial status of cooperatives.[21] Another reason was the lack of government institutions dedicated to aiding cooperatives. While the Ministry of Fomento and Public Works had a Department of Cooperatives, it lacked the means—and the will—to assist them in any way. Perhaps more troublesome, the report stressed, was that cooperatives did not emanate from the people themselves but were "imposed upon them by political parties and trade unions who planned to use the cooperatives for their own ends," a not-too-subtle jab at the Alianza Popular Revolucionaria Americana (APRA). Indeed, in the early radical days of the party, Víctor Raúl Haya de la Torre claimed that a "vast and scientific organization of a nationalized cooperative system" was essential for Peru to efficiently fight against imperialism.[22]

Despite past challenges, there was considerable optimism regarding the future. USAID stated that the "basic organization of the Indian Communities in the Peruvian Highlands lends itself to the establishment of cooperatives in the area." Furthermore, cooperatives were considered the best way to channel economic aid since fewer technical personnel could reach a greater number of people, thus accelerating social and economic development.[23] Hence, early on, traditional communities, so maligned by modernization theorists, were identified by experts as potential allies in the process of economic development. However, certain doubts remained, and on a somewhat more condescending note, the American report stressed that Peruvians did not have a sufficient knowledge of the "cooperative concept." This would be a recurring theme as American officials interacted with their Peruvian counterparts.

Interest in the development of cooperatives was not limited to Lima intellectuals and U.S. experts. New interpretations by regional intellectuals argued that cooperatives were a solution to Peru's underdevelopment, reflecting the recovery of a communal spirit lost during the Spanish conquest and now to be strengthened by the collaboration between Andean social practices and Western technology. Furthermore, intellectuals in the Junín department, particularly in Huancayo and the surrounding Mantaro Valley—a hotbed of cooperative development—saw the events taking place in the region as having great national significance.

While much has been written on the disconnect between the Mantaro Valley and Lima, existing literature has tended to ignore the synchronicity of discourse between the capital and the region in the twentieth century.[24] Take for instance Jesús Véliz Lizárraga, a Huancayo sociologist and historian who considered the central Andean region to be a "great social laboratory." An influential member of the APRA party, with whom he had a troubled relationship, Véliz Lizárraga's championing of communal development was part of an influential wider sociological analysis of Western practices in the Peruvian highlands. Véliz Lizárraga observed a trend that he judged to be a clear case of what Cuban anthropologist Fernando Ortiz called "transculturation." Ortiz had used the term to explain the encounter of Spanish and Indigenous cultures in Cuba. Véliz Lizárraga was applying the same logic to the Mantaro Valley, where a new "authentic" Peruvian man would arise out of the transculturation of "Western and Andean cultural elements." He argued that this process was also taking place in the rest of the country, but at a much slower pace than in the Mantaro Valley.[25]

In part, this was due to the historical development of the valley itself, where the colonial latifundio system had not been as dominant as in other regions of the country. In colonial times, the Spanish allowed the residents of the valley to keep small family holdings that were passed from generation to generation. Furthermore, the Spanish did not wish to radically modify land tenure, as it provided subsistence for migrant workers who worked in the mines of the central Andes.[26] Finally, after independence, the Mantaro Valley was connected to the city of Lima via the central railroad and later by a central highway, allowing it to interact politically and economically with the capital and, therefore, the coast. As such, as Norman Long and Bryan Roberts have argued in their seminal works on the central Andes, "communal traditions and patterns of community organization . . . flourished in a region that was one of the earliest and most thoroughly exposed to the effects of modern capitalism."[27] The modern and the traditional appeared to coexist in the central highlands.

This had already been evident to Véliz Lizárraga, who applauded the "exceptional" nature of the Mantaro Valley, claiming that by the 1960s the region had experienced accelerated transculturation because of the encounter of old Andean traditions and "strong republican tendencies" after independence. The main thrust of this Andean worldview was that a mentality of the collective—"of cooperation"—was needed to replace Peru's "individualist" mentality, which had been the legacy of the Spanish conquistadors. It remains quite peculiar to claim that individualist mindset had "Latin" roots. It was stranger still to associate collectivism with the Western European tradition. Yet that is precisely what Véliz Lizárraga argued. As such, he claimed that Andean collective practices were compatible with events taking place in Germany and the Nordic countries, countries that were celebrated for reaching a high level of development with a "fair distribution of wealth." Indeed, the "German Miracle," Véliz Lizárraga claimed, was the result of the cooperative mindset. Capitalist development in these regions was possible not because of Western individuality but because of Western collectivism. The new Peruvian man was no different from his European counterparts, for he also had the "spirit of cooperation and initiative." All that was needed to leave behind the legacy of the Spanish conquest was to combine the Andean collectivist spirit with Western science and technology.[28]

A more direct reference to the Incas, and the role of the state in communal development, could be found in the views of another Huancayo native, Ponciano Melgar Casallo. Although not as well known as Véliz Lizárraga, Melgar Casallo was an educator who also wrote editorials in the city's largest newspaper, *La Voz de Huancayo*. In a far more straightforward manner than Véliz Lizárraga, Melgar Casallo argued that Peru had a natural tendency toward the formation of cooperatives. Indeed, the Inca Empire was an example of "communal action and social cooperation." Thanks to this organization, the Incas had made magnificent infrastructural achievements, such as their "admirable road system and incomparable aqueducts that allowed them to cultivate the most sterile and uneven lands." Inca success was perfected as the Incas constituted "public cooperatives" under the protection of the great Inca state. These "public cooperatives" could also be found in modern Europe. Famous examples cited by Melgar Casallo include the Communal Credit Union of the Kingdom of Belgium, the Société Nationale des Chemins de Fer Français, and the British Metropolitan Water Board. These were companies of transcendental importance because they were guided by cooperative principles. In Melgar Casallo's view, Peru would have followed this course had it not been for the "arrival of the Spanish adventurers, who destroyed everything that they found because of their avarice and thirst for lucre."[29]

Melgar Casallo insisted that while this cooperative state—different from a socialist state—had been destroyed, the cooperative spirit still lived in the hearts of Indian communities. In fact, many of them "have not yet lost the vision of transforming their social, economic, cultural, and political realities through cooperative development" and could establish companies of no less importance than the great European cooperatives. Hydroelectric plants, roads, and agriculture—these were to flourish because of communal cooperative action. Melgar Casallo finishes by making a distinction between "base cooperatives"—the natural result of the existence of the Indigenous communities that made up most of the Peruvian population—and "high cooperatives," which would be represented by the supreme authorities of the state. These two types of cooperatives were the "pillars of the master cooperative edifice, so distinct from the capitalist order, leading to the socio-economic transformation of Peru."[30]

Other regional intellectuals also analyzed the growth of cooperatives in the valley. Mario Tapia, an agronomist from Puno, enthusiastically wrote in the pages of the *Voz de Huancayo* about the economic aspects of cooperative development in the Mantaro Valley. "Usually, the word cooperative has a meaning that is economic, and as such would be equated with the Incan *ayllu*, the Mexican *ejido* and the *coumvite* [sic] in Haiti." But he would also celebrate its political implications, stressing its democratic characteristics as well as its socialist ones: "Current cooperativism, is a democratic instrument, as it presents a solution to the economic and social problems of the community through the resulting harmonious collaboration of human beings and not through isolated activities—which might be antagonistic, like individual economies." Analyzing the possibility of establishing a cooperative in the valley, he stated that the existence of Indian communities made it more likely to be successful. However, cooperative values should not be attributed to the Inca Empire itself, as they were inherent in Indian communities, which guided themselves through what Tapia called "*huajite* [*sic*]", an ancient practice that had existed even before the Incas, based on the principle of "today I help you, tomorrow we help each other."[31]

Regional discourse seemed to align with Belaúnde's idea that the modern cooperative system presented Peru with the opportunity to harness what was perceived as the natural mutualist tendencies of Indian communities, this time through a Western institution. The cooperativist movement was modern, not only in the sense that it existed in the contemporary world but also because it was seen as an essential capitalist institution that had thrived in the countries of the so-called first world. These theoretical approximations would find practical expression in Peru's "new" cooperative movement, particularly in the Mantaro Valley, future site of one of Latin America's most ambitious cooperative experiments.

A VALLEY OF DARKNESS

Intellectual debates on the channeling of pre-Columbian practices through modern cooperatives were not mere abstractions. Many of them found physical expression in the creation of credit unions, agricultural cooperatives, and, in the case of the Junín department, even a communal university.[32] But the cooperative spirit—and its linkages with technological development—had found one of its earliest examples in a small town on the right bank of the Mantaro River named Muquiyauyo. The town, known as "Little Russia" because of its "progressive" spirit, had established a small communal hydroelectric plant as early as 1921. Hildebrando Castro Pozo visited the town in his capacity as director of the Indian Affairs Bureau of the Ministry of Fomento and Public Works and was mesmerized by what he considered to be the result of the mixture of the ancient ayllu and Western technology. Muquiyauyo's notoriety was also greatly enhanced by its inclusion in Peru's most famous Marxist work, Mariátegui's *Siete ensayos de interpretación de la realidad peruana*, where he celebrated the combination of Indian practices and modern technology.[33]

The Muquiyauyo communal electric company has garnered much scholarly attention, but its success obscured the fact that the rest of the valley remained in darkness.[34] Even the largest cities in the region lacked an adequate supply of electric power, a constant matter of complaint for both citizens and would-be industrialists. While cities would not be included in the plans of the future rural electric cooperative—after all, its purpose was to provide rural areas, not urban centers, with electrification—the challenges that they faced say much about the state of electric infrastructure in the area. Furthermore, many of these towns would push for the construction of small- and medium-capacity hydroelectric plants, which would eventually electrify the towns of the Mantaro Valley.

The most notable example is the city of Huancayo, the capital of the Junín department, which suffered from regular power outages. Many of these problems stemmed from the fact that the limited electric supply was meant to be a utility but also foster industrial development. The Huancayo Industrial Society, founded in 1919, controlled two small hydroelectric plants, from which the city obtained its power. These included the Ingenio plant north of the city, built in the 1930s, and Chamiseria, which had been developed and upgraded in the 1940s and 1950s. Huancayo's largest factory, the Andes Textile Factory Limited, absorbed much of the power supply, and the citizens of Huancayo complained that they were subsidizing the company through ever-increasing tariffs.[35] The Huancayo Industrial Society never had the means to increase its electric infrastructure, and rumors of expropriation surrounded the hydroelectric facilities at Ingenio, which were designed to serve the interests of a future rural electric cooperative rather than those of the city.

For the city of Jauja, situated north of the valley, the situation was even more problematic. As has been well documented, the city had a difficult relationship with the Empresa Comercializadora de Energía Eléctrica de Muquiyauyo (also known as FEBO), its main source of electric power. With no other options available, Jauja had given FEBO a fifty-year concession. Yet the company constantly struggled to adequately supply Jauja with electricity. At times, this would result in heated political battles between the two towns. Most memorably, one of Peru's first congresswomen, Jauja native María Eleonora Silva y Silva, urged her fellow citizens not to pay the electric company bills, blaming FEBO for Jauja's lack of industrial development.[36] This led the citizens of Jauja to energetically lobby the central government for the construction of a hydroelectric plant in the town of Huamalí, directly opposite Muquiyauyo across the Mantaro River. Muquiyauyo watched these developments with apprehension, jealously guarding its electric monopoly in the northern part of the valley.

If access to electric energy was problematic for the two largest cities in the valley, smaller towns such as Concepción—future site of a modest hydroelectric plant—faced even larger challenges as they attempted to develop their urban infrastructure. They were plagued by electricity shortages—the residents of Concepción complained that the service worsened even as tariffs were raised. Such complaints garnered greater attention when tied to footballing events, as was the case in a match between Alianza Lima and Vasco da Gama during the first-ever Copa Campeones de América—later known as the Libertadores Cup—in 1960. The residents of Concepción missed the match because of a power outage.[37]

While cities and larger towns had an erratic supply of electric power, the rest of the valley, with few exceptions, did not enjoy such luxuries. An editorial by the *Voz de Huancayo* lamented the lack of electricity in the small towns: "In many towns of the Mantaro Valley, this problem becomes more acute due to the rise of the population and their dreams of setting up small industries that do not emerge because of a lack of electric power."[38] There was agreement, hence, on the need to expand hydraulic sources of electricity in the region. Rural areas had been included in Manuel Prado's 1957 electrification plan, but many of the recommendations made by Électricité de France in the midfifties included small hydroelectric plants that never came to fruition, and larger ones—such as the Mantaro Hydroelectric Plant—would take far longer to come into existence than the optimistic Prado had expected.[39]

To complicate matters further, there was considerable tension between the central government and the provinces about how the expansion of electric infrastructure should be carried out, a debate spurred by the creation of the Fondo Nacional de Desarrollo Económico (FNDE). Established in 1956 the FNDE was meant to foster infrastructural development in the provinces. Its creation was due in part to a "broad provincial backlash against the economic policies of Odría, combined to favor a uniform, comprehensive approach to decentralization."[40] But although it was meant to promote decentralization, the FNDE in many ways reinforced Lima's control over other territories. This was because the provincial Public Works Committees (JOP) that the fund created had to submit their plans to the central government for approval. The Mantaro Valley was no exception, and the FNDE insisted on determining the amount of funds and their use, overriding the wishes of Junín's JOP. The problem was not a lack of funds—indeed, it was argued that the FNDE had allocated too much money to hydroelectric projects—but rather the frustration of local elites at not having a voice in such matters. The electrification of the valley, thus, highlighted the problems of Peruvian centralism and shone a light on the aspiration for greater decentralization in the rest of the country.[41]

Subsequent events proved that the central government did not necessarily know best. In addition to the existing Ingenio plant, the four principal hydroelectric projects meant to fulfill the Mantaro Valley's developmentalist dreams were those of Concepción, Huarisca, Huamalí, and Ulcumayo, located in the central and northern part of the valley. Of these four, only two became a reality. Much of the trouble can be attributed to the Franco-Peruvian consortium SOCIMPEX, which had been given the right to carry out feasibility studies nationwide. From 1960 to 1964 a parade of French engineers crisscrossed their way around the valley, submitting reports to the Ministry of Fomento and Public Works. As Alfonso Quiroz noted, SOCIMPEX became embroiled in a "little publicized indication of graft" by overbilling the Peruvian government for more than ten million dollars.[42] However, in Junín, where SOCIMPEX was in charge of building the four hydroelectric plants, the publicity that it received was far from insignificant, and no reimbursement could hide the shells of unfinished plants that served as constant reminders of the company's dishonesty.

The slow progress in the northern part of the valley became even more evident because of the apparent success of one hydroelectric project in the southern section, the Pucará hydroelectric plant.[43] Work on the Pucará project, a small-capacity hydroelectric plant, started as early as 1958, with funds obtained at the behest of local senator Manuel Alonso Martínez and Congressman Luis Sobrevilla González. Likewise, the then head of the electricity sector of the Ministry of Fomento and Public Works, engineer Fritz Vallenas, took a special interest in the project. Construction was carried out by the Italo-German-Peruvian consortium Casa Wiese. Such an arrangement, by itself, was not out of the norm. What stood out, however, was that the *comuneros* of the district, in what the press celebrated as a "noble civic gesture," had assisted in the construction of the plant to cut down costs. This made the accomplishment worthy of special attention: the bringing together of the Peruvian state, foreign expertise, and communal effort.

The plant also demonstrated the technique that would be used in small- and medium-capacity hydroelectric plants in the region. A canal was built that was almost 1000 meters long, leading to an artificial tunnel of a height of almost 40 meters, which would clean the muddy waters and create force through altitude. Although of modest height, the resulting fall, passing through the British Belton and Francis turbines, moved a 36 KVA generator.[44] The operation house and the rest of the structure were built with simple bricks. Like the construction of the plant, its inauguration represented a mix of the modern and traditional. The bishop of Huancayo, Mariano Jacinto Valdivia, representing Peru's most traditional institution, stood side by side with engineer Augusto Martinelli, subdirector of the electricity sector at the Ministry of Fomento and Public Works, representing Peru's modernist aspirations. After the ceremony, a folkloric dance contest took place, followed by a lavish banquet.[45]

The combination of the state, the Church, and the comuneros dressed in their costumes was only a superficial expression of what had just taken place. The local press stated that the plant would "alleviate the demand for this service that is yearned for by all sectors of the country, for it no longer represents only the mere satisfaction of one more need, but the industrial transformation desired by the population." The people of the region, although aided by the state, were to be championed, for "Pucará has demonstrated itself to be exceptionally gifted with communal work."[46] Well before the arrival of Cooperación Popular and the Alliance for Progress, the people of Pucará offered a blueprint of how to successfully carry out communal work in the Andes.

As *pucarinos* celebrated the completion of their new hydroelectric plant, the rest of the valley struggled to develop its electric infrastructure. The Concepción plant was finished by SOCIMPEX in 1963, although it would break down regularly and, more importantly, its modest output of 750 KW would prove insufficient to meet the long-term demands of the valley. The Huarisca project would advance slowly, and when the SOCIMPEX scandal came to light, construction was abandoned, new contractors being unwilling to take over the project. Almost a decade would pass before the project was finished. Ingenio remained embroiled in Huancayo's industrial disputes. The Ulcumayo project never got past the point of feasibility studies.

The Huamalí project was a different matter. Although it never became a reality, failure to build the plant spurred a national debate. Given its vital importance to the city of Jauja and surrounding towns, the planned hydroelectric plant was discussed in the Chamber of Deputies. APRA, hostile to Belaúnde's government, pushed for the project, which after SOCIMPEX's failure had been given new feasibility studies by the Hydrotechnic Corporation of New York. In a rare display of solidarity, members of Belaúnde's AP party echoed their demands. Deputy Mario Serrano Solís supported the project but took the debate further, mentioning that representatives of USAID had been touring the area and that the population considered the only solution to their problems to be the formation of a cooperative, an initiative supported by the U.S. experts.[47]

Such was the electric landscape when Latin America's largest electric cooperative began its operations later that year. The lack of power supply would make the first years of the Cooperativa Eléctrica Comunal del Centro difficult ones. Members were recruited, power lines erected, and meetings held. But without hydroelectric plants, there was no power flowing through the grid. The local citizenry would do their part, but hydroelectric plants required technical assistance and equipment that could be provided only by specialized companies. The cooperative, as it would turn out, would find stability only when it became connected to the Mantaro hydroelectric complex, which was being built deep in the department of Huancavelica. Yet the very same interconnectedness that made it viable would also lead to its disappearance.

A NEW DEAL UNDER THE ALLIANCE FOR PROGRESS

As the electric landscape of the Mantaro Valley remained fragmented and French engineers were on the retreat, American experts had indeed been carrying out their own studies. One of them was Francis Dimond, acting on behalf of USAID. The other was Paul Tidwell, president and manager of the Meriwether Lewis Electric Cooperative of Centerville, Tennessee. They were in charge of carrying out the first phase of studies for the establishment of a large cooperative in the region. By early 1964 Tidwell had returned to the United States, and in his place arrived Troy Mitchell, manager of the Jasper-Newton Electrical Cooperative of Kirbyville, Texas. These two men were representatives of the U.S. NRECA.

While Frank Dimond was a representative of the U.S. government, Troy Mitchell was a representative of American civil society and saw himself as an apolitical agent advocating the global cause of cooperativism. In this sense, he can be considered a "Cold War expert," a nontraditional actor whose knowledge consisted of promoting a specific type of social organization.[48] The presence of NRECA was due to a small but significant amendment to the U.S. Foreign Assistance Act of 1961. At the behest of future vice president Hubert Humphrey, the act was amended "to encourage the development and use of cooperatives, credit unions, and savings and loan associations."[49] For Humphrey—who owed much of his political career to Minnesota's farm cooperative movement—this was because the Alliance for Progress needed "a mystique all its own, capable of inspiring a following."[50] The amendment reflected Humphrey's view that the Alliance should not only be an economic effort but also have a strong ideological dimension.

This view was shared by NRECA members. Its immediate predecessor had been the Rural Electric Association (REA), established in 1935 to electrify rural areas and foster economic recovery through the modernization of farming operations. With the United States' entry into the Second World War, construction materials became scarce, and NRECA was founded in 1942 as rural Americans pooled their own resources to complete the mission of electrification. After the Cuban Revolution, NRECA's president, Clyde Ellis, had come to see rural electrification as a key component in the fight against communism and, together with USAID, began "exporting the REA pattern" to Latin America, establishing cooperatives in the Dominican Republic, Ecuador, and Nicaragua.[51] As such, rural electrification in the 1960s not only reflected the aims of the Alliance for Progress but also embodied the spirit and values of an earlier era, that of the New Deal.[52]

NRECA and USAID officials had to select an appropriate site to organize this experiment. Potential candidates included the Callejón de Huaylas in Ancash, the Sacred Valley in Cuzco, and the shanty towns that were emerging on the outskirts of Lima. But the Mantaro Valley was deemed promising for several reasons. As it was a valley, constructing an electric grid posed no significant challenges. Despite the lack of existing electric infrastructure, the generating potential of the region was considerable, as it had several hydraulic sources for the construction of small plants. The Mantaro River, which crosses the valley, had one main tributary, the Cunas River. Additionally, the Alayo River, originating from the Huaytapallana glacier, produced additional streams before leading into the Mantaro.[53] The valley was also dotted by lakes and *puquiales*, or springs.[54] Demographically, it had a high concentration of people, approximately 200,000 potential electricity users. Finally, the Servicio Cooperativo Interamericano de Producción de Alimentos (SCIPA) had been active in the area since the 1950s. Part of the aim of the SCIPA program was to industrialize food production—special attention was paid to the creation of a dairy industry—something that would be more than compatible with electrification.[55]

The Cooperativa Eléctrica Comunal del Centro was officially established on November 22, 1964, after an eight-hour discussion with members of fifty Indigenous communities of the Mantaro Valley.[56] Following NRECA guidelines, a simple bureaucratic structure was established, consisting of an administrative and a monitoring council. Eliseo Herrera Castro, from the town of Vicos, was elected to head the administrative council, while Rafael Rojas Delgado, from the town of Manzanares, would be in charge of monitoring the cooperative.[57] The cooperative would also include some very particular institutional modifications. One was the Education Committee, in charge of educating citizens in cooperative values and the use of electricity through courses, movies, conferences, and its radio program, *Pentagrama Cooperativo*. The other was the Consejo de Amautas—the Quechua title for ancient Inca wise men and teachers—where former leaders would share their knowledge with new authorities.[58] In time, the American cooperative model in Peru would acquire even more local characteristics.

Media reports in Huancayo were celebratory. *La Voz de Huancayo* proudly stated that the cooperative agreement signaled the "arrival of progress to the great Mantaro Valley."[59] This progress was represented not only by electricity itself but by the fact that the comuneros had taken matters into their own hands in developing the valley. Although not mentioning any specific plans, the Huancayo daily hoped that rural electrification would give way to the establishment of small and medium-sized industries from which large factories would inevitably arise, raising living standards in the valley to those of an industrial society. Editorials from the capital were also positive. Lima's main daily, *El Comercio*, lauded the "stimulating" nature of the project: "The valley will be completely transformed . . . It will become the second greatest commercial market, after Lima. Great opportunities [will come] for light industry and textiles in particular." The newspaper also celebrated what was perhaps an underrated consequence, which was simply to provide the "personal comfort of permanent electric lighting." It likewise highlighted the collaborative nature of the project. USAID had established the basis for the cooperative itself. The office for Cooperación Popular had put USAID and NRECA representatives in contact with Indian communities. INCOOP also sent representatives to explain the workings of a cooperative. Thanks to these efforts, Dimond stated, "it is the first cooperative in Latin America that seeks to electrify such a large area."[60]

El Comercio likewise celebrated the cooperative as a political triumph for Belaúnde and his Cooperación Popular program. Belaúnde's initiative, *El Comercio* argued, had "awakened" the people in the area, "reaffirming their sense of solidarity and mutual collaboration." Not only did Cooperación Popular prove its worth with the announcement of USAID support, but it had also "earned international prestige." USAID's support also meant the Cooperación Popular could succeed despite the existence of "political interests intent on destroying it." Belaúnde's political opponents, APRA and the Unión Nacional Odriísta (UNO)—known as the APRA-UNO coalition—had vigorously opposed the program, considering it a tool to politicize the population in favor of the government.[61]

The Cooperativa even managed to score some international coverage. "More than 200,000 residents of the densely populated Mantaro River Valley in the central Andes may soon be able to tap part of the region's vast electric power resources," stated the *New York Times*. Quoting Frank Dimond, it reiterated that the Mantaro Valley project had been selected "because of the availability of power, the density of population, and the relatively high purchasing power of the farmers, most of whom are landowners with incomes of about $350 a year."[62] The cooperative seemed poised to become the new flagship program of the Alliance for Progress in Peru.

In the press, Dimond also established the parameters of the role of USAID. Dimond stated that the USAID loan would be used to construct the grid and set up connections in homes. However, he categorically stated that USAID funds were *not* to be used in the construction of hydroelectric plants. Dimond, following the U.S. model, assumed that the plants were the responsibility of the alliance between private capital and the state, and that the cooperative was to buy energy from existing and future plants, not produce its own. He considered Concepción, Huarisca, and eventually the Mantaro Hydroelectric Plant to be potential sources of power. The newspaper shared this view, supporting the idea that the future of the Cooperativa was inevitably linked to the construction of Peru's largest hydroelectric complex.[63]

With 280 initial members and initial capital of twenty thousand soles, the cooperative had much work to do, needing to reach 5,000 members and $70,000 to be eligible for USAID funds. The communities of the valley had the assistance of the American Peace Corps (PC), who had set up a regional office in Huancayo under the direction of Marion "Tex" Ford. Overseeing over a dozen volunteers, the PC was involved in various community development projects. The volunteer assigned to the cooperative was Taras Prytula, a student from the University of Maryland. The son of Ukrainian immigrants who had been involved in the formation of cooperatives back in the old country, Prytula wished to follow in his parents' footsteps by promoting the cooperative spirit in Peru.[64] He would work closely with Troy Mitchell and Frank Dimond just two weeks after the cooperative was established. When Troy Mitchell left in 1964, Prytula aided John Taylor, the new NRECA specialist, who had formerly been employed by the Walton Electric Cooperative in Monroe, Georgia.[65]

To enlist members for the co-op, Prytula traveled to the various towns of the valley showing Depression-era NRECA movies from the back of a truck. One such movie, titled *By the People for the People*, not only showed how electricity could make agricultural activities more efficient but also emphasized the democratic dynamics of cooperative organization.[66] The movies had been dubbed into Spanish, although this was not of much help, since most of the peasantry spoke Quechua. This did not seem to matter much, for it was the first time that most of the inhabitants of the valley had been exposed to cinema, an experience they eagerly embraced. Between movies Prytula and other co-op members would present their pitch as to why residents ought to join the cooperative. The promise of a second movie was essential, as it assured that the crowd would stay through the speeches.

The actions of USAID and NRECA also had an impact at the national level. The project was so ambitious that the Ministry of Fomento and Public Works created an office of rural electrification in 1966, as part of the Directorate of Electricity and Industry. Prytula developed close links with Peruvian officials, including the head of the new division, Mario Calmet, as well as Tomás Gonzáles, a young engineer in his midtwenties who was hired as manager of the cooperative in 1965. A graduate of the National University of Engineering, Gonzáles's previous experience included working for Lima Light and Power's (LL&P) subsidiary HIDRANDINA. Together with Calmet and other specialists,

Gonzáles would travel through the United States, taking rural development courses at the universities of Michigan and Wisconsin.[67] The U.S. tour also included visiting several cooperatives in the Midwest and the southern United States.[68] Both specialists visited Prytula's parents' house in Baltimore over a weekend, where the former Ukrainian cooperativists were impressed by the amount of rice that the Peruvians requested during every meal.[69]

The experiences of another volunteer also shed light on the communal dynamics of the cooperative. William Evensen was twenty-five years old and fresh out of UCLA when, with his then wife Thea, he heeded the call of the PC. After an intense three-month training program of language immersion, Peruvian history, animal husbandry, and Outward Bound training in Puerto Rico, the Evensens landed in Lima in October 1964. Soon afterward, they traveled by train to the Mantaro Valley and finally settled in the small village of Sicaya. They moved into a one-room apartment upstairs in a two-story adobe building on the plaza. The unit had no running water but had an outhouse and, obviously, there was no electricity. Evensen was not there in any technical capacity, but rather to assist with various PC activities in the area, the electric cooperative being one of them.

In time, Sicaya would become "the best illuminated town in the valley," but at this juncture Evensen promoted the cooperative and signed up members.[70] Getting people to sign was at times difficult, not because of a lack of enthusiasm from the community, but for logistical reasons. Evensen would go door to door before dawn in an effort to catch people at home before they left for their fields.[71] Evensen would organize other activities, which he would duly write down in his journal. He screened a film about cooperativism featuring NRECA's mascot—the charismatic Willie Wiredhand—which attracted many people in town. Indeed, in time a mascot—a mix of the latter and Reddy Kilowatt with Andean clothing—would occasionally be used by the co-op in the local newspaper. Together with other members, he organized elections, placed stakes for the construction of the grid, and updated maps that would show which houses belonged to co-op members.

Peace Corps volunteers did not take long to realize that the inhabitants of the valley incorporated local elements into the cooperativist experience, in part due to what Tomás Gonzáles would dub an "accentuated traditionalism." For Gonzáles, this meant that forms of cooperative organization practiced by the ayllu were still relevant in the valley despite its early exposure to capitalism, demonstrating how traditional practices facilitated the application of the cooperative model.[72] These included the *faena*, a revival of the Inca *mink'a*, through which town residents volunteered to work on projects that were of social use for the community. All members of the cooperative were meant to donate at least two wooden poles for the electric grid as part of their faena. Evensen even participated as part of his faena, which the police—at times unaware of co-op activities—confused with an ongoing protest against the local bus service.[73]

It did not take long for Evensen to develop close friendships with those organizing the cooperative. Although very much a communal effort, it usually fell upon the most distinguished members of the town to do so, and Sicaya was no exception. The main personalities were town treasurer Ángel Navarro, prosperous farmer Rodrigo Aliaga, Eloy Barona, a high school teacher, and Óscar Baquerizo, a member of one of Sicaya's most important precolonial families.[74] The cooperative quickly developed a social life of its own. Members would celebrate patron saint festivals, organize banquets, and generally enjoy themselves. As such, in Sicaya, as in other towns of the Mantaro Valley, the organization of the electric cooperative was intertwined with the celebration of communal life, and, indeed, the latter reinforced the bonds of the former.

Social customs were essential not only for the working of the cooperative but also to incorporate foreign actors. For instance, PC volunteers were particularly impressed with the ability of the locals to drink and socialize regardless of the time of day, a common social custom in the Andes. Evensen remembered one instance in which, after a long morning of knocking on doors with city treasurer Ángel Navarro—at over two hundred pounds, "the largest man in town"—Ángel ordered both men a liquid breakfast at the corner bodega, a few *copitas de aguardiente*, before their eight o'clock meeting with electrical engineers. The meeting was a big success, for "both engineers and locals were pleased with the progress of the electric co-op and the number of families buying shares." Such progress called for a celebration. "A couple of bottles of *pisco* suddenly materialized, a bottle of champagne followed, a case of quarts." By 10:00 a.m. Evensen was "blind drunk," and the group had moved out to the plaza, where they continued "drinking, telling jokes, talking about the co-op, and local crops." By 11:30 a.m. Evensen left the group to go to his house, where he collapsed on his bed. He woke up in the late afternoon. The room was cold and dark. Hating to light his oft-malfunctioning kerosene lantern, Evensen went out onto the plaza to buy a candle at the bodega. Inside the bodega he found Ángel Navarro seated with three other men. They were still drinking. As Evensen recalls, "as soon as I walked in the dirt-floored room, Ángel turned toward the bodega owner, and said '*Señora, por favor otra*

FIG. 13. Evensen (back row, right) and fellow *sicainos*.
William Evensen, private collection.

botella, nuestro amigo ha regresado.' (Another bottle please, our friend has returned). To which the bodega erupted in cheers." It was the "gringo's" turn to buy.[75] Such experiences meant that Peace Corps volunteers approached the dynamics of the cooperative not from a purely technical standpoint but also from a social one, in which human relations were key.

Evensen clearly remembers when president Belaúnde visited the city of Huancayo in November 1964. The cooperative was in its infancy, but its members were eager to show their support—and existence—to the president. At the end of the parade, the members posed for a picture, giving William and Thea the honor of being at the center of the photograph. Yet, when the couple left in 1966, they did so without any knowledge of whether the cooperative would be a success or not. Indeed, there was still no electricity flowing through Sicaya. However, both had become part of the community. Decades later, the family of Óscar Baquerizo would recall that the head of their eminent Sicaya family fondly remembered the "gringo" until the day he passed away.

The cooperative quickly acquired the "mystique" that Hubert Humphrey sought. By April 1965 it had gathered 2,500 members, aided by a newspaper and radio campaign directed from its central office, located in El Tambo district, close to downtown Huancayo. By midyear, the cooperative was well on its way to reaching five thousand members, enough to qualify for USAID funds. However, despite the evident enthusiasm of the communities in the Mantaro Valley, the Cooperativa stopped registering members that same year.[76] The USAID funds were simply not arriving. A dispute between Belaúnde and the United States—ironically, over another source of energy—threatened to snuff out the life of the cooperative before it could effectively come into existence. This dispute was over the potential expropriation of the U.S.-owned International Petroleum Company (IPC). To force a settlement that would favor the IPC, the U.S. government froze all USAID funds to Peru. This infuriated NRECA and the PC volunteers involved in the cooperative, and, of course, all the Peruvians taking part in the project.

Cooperative members took matters into their own hands, and after speaking to PC volunteers in the area, they decided to organize a delegation to travel to Lima to speak to USAID representatives.[77] In November 1965 a group of thirty-four members met with Frank Dimond, who, without explanation, told them that the funds were not available. After contacting AP deputy Víctor Alfaro de la Peña, they traveled with Mario Calmet and INCOOP director Víctor Camacho to plead their case to fomento minister Sixto Gutiérrez and President Belaúnde himself. Shortly afterward, Belaúnde contacted American ambassador Wesley Jones and told him that if the loan was not approved for the project, all PC volunteers would be expelled from the country.[78]

A few months before the delegation visited Lima, a dramatic development was taking place in eastern Junín that would also influence the course of events. In June 1965 Luis de la Puente Uceda and his Revolutionary Left Movement (MIR) initiated guerrilla activities east of the Mantaro Valley. Newspaper reports—certainly exaggerated—stated that MIR had created the "Socialist Republic of Pucutá" deep in the jungles of eastern Junín, possessed sophisticated armament, and were being trained by foreign elements. There were unsubstantiated rumors that Ernesto "Che" Guevara might even be involved.[79] Finally, higher authorities intervened, although for less than altruistic reasons. Ambassador Wesley Jones tried to persuade Jack Vaughn, assistant secretary of state for Latin America, that a small loan of $1.6 million could positively affect the outcome of the dispute, as it would demonstrate to Peru "the advantages of concluding the IPC negotiations successfully."[80] The money was slow in coming; it would be approved only after President Lyndon B. Johnson had lifted the ban on loans to Peru, along with the personal intervention of U.S. national security adviser Walt Rostow.[81] The specter of communism in the area did not make Washington think twice about supporting the IPC, but it does seem to have nudged the U.S. State Department to rethink approval of the loan to the electric cooperative, as this region was shaping up to be the main front of Peru's Cold War.

The loan finally arrived in January 1967. The $1.6 million from USAID was complemented by a loan of $600,000 from the Peruvian government. The Cooperativa also contributed almost $50,000, an extraordinary amount given the purchasing capacity of the residents of the valley. Also extraordinary was the fact that those who became members were gambling on the success of a project still in its infancy. One consequence of the arrival of USAID funds in the valley was that Muquiyauyo's dreams of expansion were effectively dashed. After failing to obtain money from a British consortium, FEBO representatives contacted USAID for a potential loan. However, Francis Dimond stated that granting such a loan was out of the question. According to the logic of both USAID and NRECA, if they were to give a loan to Muquiyauyo, then any other small town in the valley could directly ask for funds.[82] This was to be a regional cooperative. Electrification of the region would happen on a grand scale, or it would not happen at all.

The release of the funds inevitably led to another dispute: who should the funds be given to? USAID felt more comfortable dealing with the Peruvian government, but NRECA argued that the funds should be given directly to its fellow cooperative, demonstrating that NRECA experts did not consider themselves political agents. The former option was also opposed by the cooperative itself, and the media openly spoke of tensions between the cooperative and the Ministry of Fomento and Public Works. Indeed, Huancayo's only newspaper sided with the cooperative, arguing that if given to the government the funds would be used for "bureaucratic tourism."[83] For Huancayo's elites, the cooperative, even if funded and influenced by an external player like the United States, had greater legitimacy than the Peruvian central government. This government, although seen as a distant entity, was considered a threat to the efficient working of the cooperative. In the end, NRECA would give technical assistance directly to the cooperative, while USAID would deal with the government. This arrangement proved to be useful: when the arrival of USAID funds became uncertain because of tensions between the two countries, technical assistance from NRECA remained in place.

Within NRECA, two specific cooperatives assisted with the project: the Arkansas and Texas electric cooperatives. Rather than sending experts to the south, Peruvian engineers traveled north, disrupting conventional patterns of knowledge flow during the Cold War. Both cooperatives provided training in the form of hosting visits to the United States, as well as providing surplus material. Regular two-month training stints were offered by the Arkansas Electric Cooperative in Little Rock. The visits of young Peruvian engineers were celebrated in the local press, as they toured different sites in the state, such as the Ouachita Electric Plant.[84] Their personnel were active in Peru as well (indeed, posters produced by the Cooperativa would proudly add "with the assistance of the Arkansas Electric Cooperatives"). As for the Texas Electric Cooperative, it offered surplus materials to the Cooperativa del Centro. However, sharing this technically sophisticated material presented challenges. Surplus meters, for instance, would have to be converted to Peruvian voltage, which meant that the costs would be very high. Peruvians could, however, hire a meter technician who was trained in conversion, which would reduce the costs and, as an added bonus, "would have the advantage of furnishing a job for someone."[85] Correspondence between NRECA and the Texas cooperative showed genuine interest in the Peruvian fellow cooperative, stressing that "any way that we could help to reduce the cost would help extend the service further to the thousands who are in need of it."[86]

By modernizing their grids and sending surplus materials to Peru, American cooperatives ensured that the spirit of the New Deal was present not only in the personal connections developed among institutions and their officials and in the REA-era manuals circulating in the valley but also in sharing the very equipment that had once powered rural America. While the cooperative was grateful for the assistance, it was concerned about developing a dependency on American materials. Given the troubled relationship between the two governments, this caused some anxiety. Gonzáles, for instance, argued that in the future the cooperative should acquire domestic equipment, and even develop close links with other cooperatives in Latin America.[87]

This north-south technological exchange also impacted the environmental realities of the valley. NRECA officials celebrated the fact that the Ministry of Fomento and Public Works, despite some apprehension, had accepted the idea of using wooden poles to connect the towns, a common practice in most rural electrification endeavors. This decision lent itself to some curious experiments. SCIPA had been growing eucalyptus trees in the town of Apata, in the northern part of the valley. Here was a possibility for the two programs to complement each other. Samples of the *Eucalyptus globulus* trees were sent from Peru to Minnesota for testing, in the belief that "they could work well if taken proper care in the central highlands of Peru."[88] Unfortunately, the wooden eucalyptus poles rapidly deteriorated, and had to be replaced with sturdier poles made of pine trees. Despite this upset, the two-line wooden poles became a fixture in the valley. If one were to look up at the sky and see only the electric grid, one would not know if they were standing in the U.S. Midwest or the central Andes.

It appears that the cooperative had an immediate impact not only on the daily lives of the inhabitants of the Mantaro Valley but also on the ways that electric imaginaries were created, and Andean imaginaries of electricity proved to be as strong as those in the American West when it was electrified.[89] A survey carried out by Billy L. Ramsey, a specialist sent by NRECA in July 1968, sheds some light on the subject.[90] The survey shows that electricity was mostly used for lighting, although all of those interviewed stated that they were planning to purchase electric goods. The most sought after were electric hot plates, food blenders, radios, and sewing machines. (One interviewee even looked forward to purchasing a projector). All of the thirty people who were interviewed throughout the valley were aware that electric services were being provided by the cooperative, and, when asked who "owned" the cooperative, the majority answered that the members were the actual owners of the electric grid. But while it was clear that the co-op members were aware that their cooperative "belonged" to them and had many ideas about what the cooperative should do to improve their lives, it seems that they were not quite sure what they could do for the cooperative. Indeed, when Ramsey asked this complex question, people did not understand it, and the only response he obtained was from one member who proudly answered that "he was going to help keep the children from breaking the streetlights and report any problem to the cooperative office."

This confusion did not mean that the people of the area did not feel like active members of the cooperative. Most of them proudly stated that they had attended meetings and would continue to do so in the future. While there was some confusion as to where the money had come from to finance the cooperative—some said the United States, others the Peruvian government, and a few more believed that it came from the members themselves, all correct answers—those interviewed concurred that a cooperative, and not another type of organization, best served their electric needs.

Some five years after Ramsay's survey, Tomás Gonzáles also believed that the cooperative had fundamentally altered the lives of the residents. While appliances were slow in arriving, they were already present in the valley by the 1970s. Electricity had also impacted educational dynamics, as students could do their homework after dark. Socially, electricity had created

FIG. 14. Electricity arrives at the town of San Jerónimo, Mantaro Valley. Perú. Presidente, *El Perú construye: mensaje presentado al Congreso Nacional por el Presidente Constitucional de la República, Arquitecto Fernando Belaúnde Terry, el 28 de julio de 1968* (Lima: Minerva, 1968). Biblioteca Nacional del Perú.

a greater sense of "social cohesion," putting to rest fears that electrification might lead to greater individualism, as parks, plazas, and recreation centers were now frequented at night, thanks to public lighting.[91] While electrification was certainly used as a public utility, it was also meant to foster industrial development, and in this field the results were mixed. Plenty of imaginaries abounded as to what industries could be electrified, among them grain mills, lumberyards, sausage makers, textiles, and even an ice cream factory. Yet it wasn't until the 1970s that the Tupac Amaru cattle cooperative and the Concepción dairy plant—the latter "the true arrival of industrialization"—benefited from the electricity created by the cooperative.[92] Yet, despite not resulting in an industrial revolution, it was undeniable that electricity was slowly changing the economic and social characteristics of the Mantaro Valley.

COOPERATING WITH THE REVOLUTION

As the cooperative grew, Belaúnde's government was in the midst of a political crisis. Programs like Cooperación Popular had showed some tangible results, but his other attempts at reform—particularly agrarian reform—had been blocked at every turn by a hostile Congress. Likewise, the IPC dispute had frozen much of the Alliance for Progress funds destined for Peru. Because of these difficulties, Belaúnde told U.S. officials that he was disillusioned that his government had been unable to deliver the reformist promises made during his campaign.[93]

This disappointment was shared by the Peruvian armed forces, who deposed Belaúnde in October 1968. The new government, self-proclaimed as the Revolutionary Government of the Armed Forces (GRFA), had far more ambitious plans than its military predecessors. Historically, the armed forces had always intervened on behalf of the Peruvian elite, but during the Cold War they realized that supporting Peru's wealthier classes did not promise the same stability that it once had. A new type of thinking developed within the institution that would become known as "integral defense." If the country was to be spared from a communist revolution—like the one attempted in eastern Junín, which had had a profound impact on the armed forces—then it was necessary to eliminate structural sources of discontent, in other words, underdevelopment. To break the dependent nature of the Peruvian economy, the revolutionary government expropriated large agricultural estates and distributed them among the peasantry in the form of agrarian cooperatives. Furthermore, it embarked on an ambitious program of Import Substitution

Industrialization (ISI), of which the generation of electric power became an integral part. The cooperative would now have to deal with new realities as the new government sought to establish a nonaligned policy and move closer to the Eastern bloc.

By the time the military government took over, the cooperative had ten thousand members and had placed infrastructure in thirty-five districts in central Junín and come into contact with over a hundred communities. Furthermore, the Cooperativa continued to strengthen its transnational links. The cooperative hired the U.S. engineering firm Stanley Consultants of Iowa to plan the construction of the grid on the right bank of the river. The U.S. consultants worked side by side with the emerging Peruvian engineering firm Piazza y Valdez, founded by Walter Piazza Tanguis and José F. Valdez Calle.[94] Despite the impressive numbers, it was evident that the Cooperativa had grown faster than it could provide electricity to its members. Its services were still limited to the left bank of the river, where its modest Concepción and Ingenio plants were located. The combined output by both plants was not enough to power an area of over 1,000 square kilometers through the 500 KM of electric lines that crossed it. Thus, electricity continued to coexist with traditional candles and kerosene lamps. What was needed was a medium-capacity plant on the right bank of the river.

The cooperative-based "revolution" was at first welcomed by the Cooperativa del Centro, although there was considerable anxiety as to how relations would develop. A delegation traveled to Lima to meet with General Jorge Fernández Maldonado, head of the newly created Ministry of Energy and Mines, the old Ministry of Fomento and Public Works having been dissolved. The minister would not meet with them at first, as authorities considered the project to be another one of Belaúnde's "hijinks."[95] The delegation then traveled to the U.S. embassy, where American officials informed them that the cooperative would continue to be funded. Indeed, that same year, a propaganda movie called *Light of the Andes*, in which the United States touted its hydroelectric projects in Peru, was released. It included a brief segment of members of the co-op voting in an assembly. This proved that there was still interest in the project.[96] Furthermore, NRECA, which saw itself as a partner of the cooperative rather than the government, continued to provide technical assistance through its Arkansas and Texas members.[97] In time, however, as the United States and GRFA sparred over a number of issues—including, but not limited to, the IPC problem—the American presence would start to fade away.

Such events were still to come, however. In the meantime, the co-op members sought to win favor with the new government. As the cooperative's fifth anniversary approached, and knowing the new regime's tendency toward cooperative organization, it decided to hold a "Cooperativist Folkloric Parade." The parade took place in downtown Huancayo on November 22, 1969, and included forty-three towns that had joined the cooperative over the past five years. Hundreds of cooperative members marched downtown using the cooperative salute—like the Olympic salute—and referred to one another as *hermanos cooperativistas*. More significantly, each town performed their *estampa costumbrista*, which depicted various historical and communal events.[98] These included *La Toma del Inca* (the capture of the Inca by the Spanish), the famous dance of the Negritos—horses and riders included—the *Avelinos de Cáceres*, celebrating the resistance of the Peruvian marshal in the Mantaro Valley during the War of the Pacific, and other themes such as the sowing and harvesting of the land. It appeared as if local customs and the cooperative spirit had achieved complete synergy.[99]

The parade proved to be a success, and extensive media coverage followed. Soon, Gen. Fernández Maldonado contacted the cooperative, which once again sent a delegation to Lima. A public appeal was made to the revolutionary government to finish the Huarisca plant, abandoned since the days SOCIMPEX had been active in the valley. The government took control of the project through its Servicios Eléctricos Nacionales. By midyear the long-awaited project was underway. The inauguration of the "new" plant was a momentous event. Carlos Mariotti, head of LL&P, was present, further cementing the company's new social policy toward rural electrification. More importantly, Gen. Fernández Maldonado visited Huancayo and was a guest of honor of the cooperative during another colorful parade, in the town of Sicaya, home to the Huarisca plant. The minister also toured other towns in the valley, where he received a joyous reception. The cooperative took out a full-page ad in the local newspaper welcoming their "fellow cooperative brother, General Jorge Fernández," and the "pioneers of rural electrification of the fatherland" expressed their solidarity with the principles of the Peruvian revolution by "confirming our indefatigable decision to contribute to the accelerated development that is taking place in all sectors of the country."[100]

If the Cooperativa sought the support of the armed forces, the armed forces likewise portrayed the success of the Cooperativa as the success of the revolution they had undertaken. Another ad in *La Voz de Huancayo* depicted a huge electric tower next to a peasant in traditional Andean garb with a raised fist. The caption above the ad read, "Light comes to the peasants with the revolution." (See figure 16.) Perhaps more importantly, the government established a clear link between rural electrification and its program of agrarian reform. General Fernández stated in his speech that "it is not enough for the peasant to own the land he works, but this must be complemented with rural electrification not only to foster economic growth, but also to achieve the comfort and happiness that he is entitled to."[101]

Despite this promising start, relations between the military government and the cooperative soured in the following years. It seems the armed forces simply did not take kindly to the cooperative's autonomy, as it predated the cooperatives that had been established by the revolutionary government. In 1972 the government announced that the cooperative would be audited—by the new institution in charge of cooperative development, the Oficina Nacional de Desarrollo Cooperativo (ONDECOOP)—over the possible embezzlement of funds. The leadership of the cooperative argued that the audit was the result of a campaign of "lies and defamation" being spearheaded by members who had been denied special privileges. One such member was Eliseo Herrera Castro, the cooperative's first head of the administrative committee, who had been replaced in 1966 due to an alleged improper use of the cooperative assets, including using its only vehicle to campaign for the Huancayo city council on the then APRA-UNO ticket.[102] Likewise, it claimed that all the attacks had been encouraged by the regional representative of ONDECOOP. The directors needed to tread carefully with this last accusation. The cooperative once again expressed its loyalty to the supreme revolutionary government as a champion of the cooperative movement but denounced its local representative as a sworn enemy of the system.[103]

A five-hour assembly was held, where heated exchanges took place between the Cooperativa and members of ONDECOOP.[104] More worrisome, representatives of the Sistema Nacional de Apoyo a la Movilización Social (SINAMOS)—dressed as civilians, acting as the "eyes and ears" of the revolutionary government—appeared during the assembly. The audit found some irregularities but did not consider them to be voluntary omissions, and manager Tomás Gonzáles stated that while there had been some errors in the financial books for the 1970–71 period, these had already been rectified.[105] Despite being absolved and having the support of the majority of its members, Gonzáles would cease to be the cooperative's manager. He presented his letter of resignation when Eliseo Herrera Castro once more became head of the administrative council after the audit.[106]

FIG. 15. Sicaya-Huarisca Power Plant. Perú. Presidente, *El Perú construye: mensaje presentado al Congreso Nacional por el Presidente Constitucional de la República, Arquitecto Fernando Belaúnde Terry, el 28 de julio de 1968* (Lima: Minerva, 1968). Biblioteca Nacional del Perú.

FIG. 16. "Light comes to the peasants with the revolution." *La Voz de Huancayo*, June 23, 1970. Biblioteca Municipal Alejandro Deústua.

The cooperative faced another challenge when the revolutionary government passed a new law of electricity that same year. The law stated that "electric power is present in almost all productive activities, as well as [being] a good that must be available to the collective. The supply of electricity for the public is essential for the economic and social development of the country, and it also constitutes a strategic instrument."[107] Thus, according to the armed forces, electric power was vital to revolutionize the country, and while other sectors of the economy might organize themselves in a cooperative fashion (indeed, the government promoted such organization) electricity would thenceforth be generated, transmitted, and distributed by the state.

It took some time for the members of the cooperative to understand the true consequences of the new legislation. The law would be applied gradually, and the military government assured private companies that it did not include expropriation. In any case, the government could take no such action until it could provide the region with an adequate supply of electricity, which did not happen until 1973, when the Mantaro Hydroelectric Plant was inaugurated downstream. Even then, it would take a few more years for the government to feel confident enough to nationalize all electric services. In the meantime, with the first phase of the Mantaro plant in operation and producing over 300 MW, the cooperative now had access to an almost unlimited supply of energy. More importantly, when connected to the Mantaro Hydroelectric System, it ceased to be an isolated cooperative in the central Andes; it was now connected to a greater national grid.

In 1975 the Cooperativa had to deal with a more dramatic turn of events. Gen. Juan Velasco Alvarado, the head of the revolutionary government, was displaced by Gen. Francisco Morales Bermúdez. During his last months in office, Velasco had radicalized government policy, amplifying the existing divisions within the armed forces. The arrival of Morales Bermúdez—a member of the moderate faction—did not signify an immediate change; indeed, during his first months in office Morales Bermúdez announced that the government was committed to pursuing a path toward "Peruvian socialism," as he needed to buy time before putting an end to the most radical aspects of the revolution. Nevertheless, his words had real consequences. A damning editorial in *La Voz de Huancayo*—which a decade earlier had championed cooperative development as a road to capitalism—now condemned it for the very same reason. "The cooperativism that we know today, be it of products or services, has been left behind by history."[108]

This meant that traditional cooperatives were no longer desired; rather, the revolution now required the creation of "social property" cooperatives. The traditional capitalist cooperative was a for-profit enterprise and only benefited those who had acquired shares. Thus it offered, at best, only a limited solution to the structural problems of capitalism. In contrast, social property cooperatives were destined to serve the interests of all members of society, not only shareholders. Furthermore, social property cooperatives would increase worker participation in decision-making. The Cooperativa del Centro, which predated the revolution, was an obvious target. Created under the NRECA model, it was inevitably associated with the most capitalist of countries. In any case, NRECA representatives and PC volunteers were long gone as U.S.-Peru relations had been steadily deteriorating since the 1968 coup, and newly arrived Hungarian technicians walked through the halls of the local university while the Russian ambassador toured the valley.[109] This inevitably affected personal relations as well, and Taras Prytula sadly recalls that after 1968 the letters between him and his friends from the defunct Ministry of Fomento and Public Works "simply stopped coming."[110]

The association between socialism and cooperativism was resented by many working for the cooperative. Luis Carlos Arroyo, a young engineer from Huancayo who had graduated from the Universidad del Centro, considered the association to be detrimental to the cooperative movement in general, a reality that became more evident when the military "experiment" came to an end. Faithfully carrying a copy of NRECA president Clyde Ellis's *A Giant Step*, Arroyo made it clear that he identified with the original values of the cooperative—no doubt reinforced by his training stint in Arkansas—and resented the intromission of the government, which he claimed, "had no idea of what we were doing here most of the time."[111]

In 1976 relations between the government and the cooperative reached a sour conclusion. A loan of fifty-three million soles had been given to the Cooperativa to expand its activities, yet it had been unable to expand those activities or to repay the loan. It remains unclear where these funds came from, as the loans made during the Belaúnde government were due in thirty-five years.[112] While in the early years of the revolution the government might have been lenient toward the cooperative, it would be lenient no longer. That same year, the government canceled the territorial concession given to the cooperative and demanded that all its infrastructure be handed over.

The cooperative did not wish to be absorbed by the government. Indeed, during the general assembly of 1976, both its directors and its members objected to the government's actions. When it came to terms with the inevitability of expropriation, the Cooperativa sought to find another reason for its existence. Plans for entering the cement business, or becoming involved in the fertilizer industry, were considered. On a nostalgic note, the cooperative continued to oversee the construction of wooden posts, before it silently faded away.[113]

CONCLUSION

The Eléctrica Comunal del Centro Ltda. No. 127 represented a moment, in Peruvian history, of belief that social and economic development could be achieved by the creation of a hybrid discourse that merged Peruvian traditional practices with capitalist organizational models, not solely by the wholesale importation of modernization blueprints from the so-called first world. Such beliefs found acceptance not only among regional and national intellectuals but also in U.S. foreign aid agencies, who considered collaboration and investment in Indigenous communities to be compatible with cooperative development. However, given the ambiguous ideological interpretations of cooperativism, the association of early cooperatives with capitalism proved to be detrimental in the volatile political context of Peru's Cold War, as the global conflict was fought through infrastructural development.

Despite its troubled history, the cooperative successfully electrified over one hundred population centers in the Mantaro Valley. But it was as much a political as a technological process. Conflict and tension—both domestic and international—highlighted how the politicized nature of Peruvian electrification impacted not only urban centers but rural populations as well. When the state seemed distant during Belaúnde's first term in office, local communities could determine how the business of the cooperative was to be conducted, even if this autonomy meant political conflict within the institution. With the arrival of the GRFA, this autonomy disappeared, and relations between the state and local actors deteriorated. Regional infrastructural development was no longer to be determined by the residents of the valley but by a strengthened central government. This process reached its natural conclusion when the cooperative was connected to a national grid, leading to its dissolution.

Cooperative development in central Peru also demonstrated how scientific approaches toward social organizations emanating from the first world quickly acquired local characteristics. This was manifested not only in the bureaucratic structures of the cooperative but in the daily lives of cooperativistas. The latter incorporated local practices and customs, ranging from existing communal labor practices to cultural expressions, as the residents of the Mantaro Valley inevitably expressed their own traditions when interacting with the models that transnational experts brought to Peru. Local knowledge and practices, hence, were by no means lost in the quest for development, and tradition was no impediment to economic modernization—in fact, it proved to be an asset.

By 1979 the Peruvian armed forces had returned to the barracks, and in 1980 Belaúnde returned to power after winning the election that same year. After the so-called military revolution, there was not much left to reform, and Belaúnde would undo many of the radical changes that had been pursued by the armed forces. Cooperatives were dismantled and turned into public limited companies. Once hailed as pillars of communal capitalist development, the association of cooperatives with the military government and their "socialist" overtones proved to be detrimental in subsequent years. Furthermore, the electricity sector was privatized, signaling the arrival of neoliberalism in Peru. All that remained was a law passed by Belaúnde, during his second term, decreeing that a "National Day of Rural Electrification" was to be celebrated every November 22, the day that the cooperative had come into existence.[114]

5

Electric Revolutions

In August of 1964 Elsa Marlene Castellares claimed that the "Sun of Mantaro" would rise in the central Andean city of Huancayo, "the geographic heart of our country." Castellares was a young member of Acción Popular (AP), a political party founded with the aim of recovering pre-Columbian grandeur via infrastructure projects. In this opinion piece in her local newspaper, she went on to establish an analogy with the foundational myth of the Tahuantinsuyo, or Inca Empire. Castellares noted that "just as Lake Titicaca was the legendary *paqarina* of the glorious *Tahuantinsuyo*, Lake Chinchaycocha will be the site of our economic emancipation." The *paqarina*—a Quechua word meaning "sacred place of origin or birth"—of Lake Chinchaycocha, or Lake Junín, was the source of the Mantaro River, where Peru's most ambitious hydroelectric project was to emerge. Castellares lamented that Peru did not control its oil, its mines, even its ocean, but now the "sad hoofprints of the horses of the *Conquista* would be erased forever by the horses of Mantaro." Castellares concluded that Mantaro "*vale un Perú*."[1]

This chapter traces the construction of the Mantaro Hydroelectric Plant, arguably Peru's most ambitious infrastructural project of the twentieth century. Negotiated and built over three political administrations, the social and economic significance of the Mantaro Plant reflected changes in transnational ideas of development. Established as a public-private partnership during Manuel Prado's second government, the project was informed by Modernization Theory, with the aim of accelerating industrialization and weakening communism, resulting in a "right kind of revolution."[2] However, Peruvian interpretations of Modernization Theory also interacted with previous local infrastructural legacies of the nineteenth and twentieth centuries, giving way to peculiar institutional arrangements. This phenomenon has often been overlooked in previous studies of the Global South, as they focus mostly on newly independent states that emerged after the wave of decolonization in the second half of the twentieth century, unlike Peru and other Latin American nations that had a longer existence as independent political entities.[3]

The Mantaro "scheme" subsequently faced political difficulties during the *accion populista* government of Fernando Belaúnde Terry, whose ambivalence toward the project was used by his opponents—both by those who supported and opposed it—to either weaken or strengthen the role of the state in economic development. The plant was finally inaugurated by the Revolutionary Government of the Armed Forces (GRFA), led by General Juan Velasco Alvarado, who wished to carry out a "revolution from above." While sharing their predecessors' fear of communism, the new "revolutionary" government dialogued with Dependency Theorists, for whom Peru's economic backwardness was the direct result of the prosperity of first-world nations, turning the Mantaro project into a tool against underdevelopment. One single infrastructural project, hence, allows one to trace changing perceptions of development almost within a single decade, complicating existing understandings of infrastructural imaginaries.

The Mantaro project was not limited to political controversy and infighting. It was also a massive engineering achievement in which Western scientific knowledge was applied to a distinct Andean topography by Peru's most famous engineer, Santiago Antúnez de Mayolo. Trained in Europe and the United States, Antúnez de Mayolo mixed Western science with local knowledge, representing a case of scientific excellence from the periphery.[4] Antúnez de Mayolo's involvement was far from being only of a technical nature, as he became politically committed to the project. This commitment became essential as German, British, and Italian firms fought one another to finance the Mantaro Plant.

Finally, the Mantaro project encapsulates Peru's political battles over the country's decentralization, showing the importance of infrastructural politics in the urban-rural divide.[5] While the Mantaro Plant was meant to industrialize the countryside, Lima elites insisted on maintaining total control over the project, much to the chagrin of regional elites. The stakes were considerable. For the threatened Peruvian oligarchy, the industrial revolution promised by Mantaro would render agrarian reform irrelevant, as inhabitants would flock to fill new industrial jobs. Likewise, Prado hoped that industrialization would halt Andean migration to Lima, once again preserving colonial imaginaries of the capital, as electricity could be used to maintain traditional racial divisions. However, when the GRFA took over the project, it was depicted as a tool to end Lima's oligarchic dominance. The Mataro plant, thus, allows one to explore Andean electrification in a way that complements the various studies of that "peculiar" regime.[6] In the end, neither oligarchic nor "revolutionary" goals were fulfilled. Because construction took well over a decade—and due to the uncertain assumption that the project would lead to rapid industrial development—migrants continued to race to the capital. By the first half of the 1970s, the city's population had grown to over three million inhabitants, reinforcing its political and economic importance.[7] In a final paradox, Lima's shantytowns were powered by the electricity generated by the distant Mantaro.

ALL THE GOLD OF THE CONQUEST

In June of 1901 Nemesio Augusto Ráez y Gómez, subprefect of Tayacaja, a remote province between the Huancavelica and Junín departments, set off from the city of Pampas to explore the Mantaro River and its tributaries. Ráez y Gómez was following the government's mandate that officials explore surrounding territories with the aim of furthering the state's knowledge of distant lands. The goal of the expedition was to reconnoiter the jungle region east of the province. Unlike earlier explorers, who shrouded the region in "fables and legends," Ráez y Gómez claimed he departed with a "mandate of science." He sought the best route from his province to a suitable river port, to provide a commercial outlet for the nearby departments of Junín, Huancavelica, and even Ayacucho. The success of the expedition hinged on time, as Ráez y Gómez had limited means at his disposal. Furthermore, he could not neglect his political responsibilities in Tayacaja. Armed with a few rudimentary scientific tools—a chronometer, a thermometer, a compass, a spotting scope, and a theater glass—Ráez y Gómez embarked on his journey with his secretary Julio Gamarra and two police officers. Heading northeast, Ráez y Gómez was following in the footsteps of Antonio Raimondi, who decades earlier had attempted to find the river mouth of the Mantaro.[8]

Soon, other "notables" joined the expedition. As the ever-growing band of explorers pushed eastward, they encountered lost ruins. An abandoned factory with large ovens, offices, and a chapel, as well as the vestiges of *andenes*—ancient, stepped terraces carved into the mountains by the Incas for cultivation—was evidence that the area had thrived in the past. When the subprefect reached his destination, he found himself in the Amazon basin, surrounded by the "deafening" sound of the roaring Mantaro, which contrasted with the silence of the forest. Given the river's violence, Ráez y Gómez thought that its greatest potential would be motor power, which could be transmitted from this isolated region, through electric wiring, to power faraway lands. The problem, as Ráez y Gómez noted, was that the state was completely absent from this remote territory. "The whim of nature, to have planted such greatness and majesty in places where the hands of men, perhaps, will never be able to use them!"[9]

It would take four decades before a second expedition explored the hydroelectric potential of the Mantaro River. This time, an experienced engineer named Santiago Antúnez de Mayolo would reconnoiter the site. In time Antúnez de Mayolo would become the country's best known engineer, largely because of his ability to selectively appropriate foreign scientific models and adapt them to an Andean landscape. As such, he fell within the parameters of what Stuart McCook has described as a "creole scientist," "creole" in the sense of hybrid, but at the same time local. Likewise, through his hydraulic knowledge, he represented a new expression of *saberes andinos*, and his future hydroelectric endeavors a clear instance of scientific excellence carried out in "peripheral" latitudes.[10]

Born in 1887 in the department of Ancash in northern Peru, Antúnez de Mayolo belonged to an old colonial family of the Aija district. The future engineer was a proud *provinciano*, and, despite Ancash's proximity to the sea, Aija was distinctly Andean. He had great pride in his family's accomplishments, particularly those related to infrastructure. After the War of the Pacific, when the state was absent in the provinces, his father and other notables organized the construction of a road between Aija and the Bellavista province. In his memoirs, Antúnez de Mayolo remarked that "looking at that road, I developed the audacity of not considering the impossibility of carrying out the great engineering projects that I would elaborate."[11]

Antúnez de Mayolo's interests were not limited to engineering, as he was also fascinated with the history of Peru's ancient civilizations. Aija was an important archaeological center, and early in his childhood he learned the names of his hometown's different neighborhoods, which in Quechua corresponded to the different parts of a llama or possibly a vicuña. Later, as an engineer, Antúnez de Mayolo continued to study archaeological ruins and became an expert on pre-Columbian civilizations in his own right. He led several expeditions to the pre-Incan city of Chavín, becoming obsessed with the symbolic meaning of the city's monolith. He would eventually be named by the government as head of other official excavations.[12] Paradoxically, the man of science seemed to be deeply interested in the mythological histories of the ancient.

Antúnez de Mayolo attended Catholic school in Aija, before transferring to Lima's famous Colegio Nacional Nuestra Señora de Guadalupe for his secondary studies. He enrolled at San Marcos University in 1905 to study mathematics and engineering under notable professors such as Federico Villareal and Emilio Guarini. In 1907, under Guarini's advice, he departed for France, where he attended the Grenoble Polytechnic Institute. Guarini told Antúnez de Mayolo that he would be in a "region rich in waterfalls just like Peru, and because of it, heavily industrialized."[13] In Grenoble, he studied under Louis Barbillion and George Flusin, the latter particularly well known for his research on the drilling of glaciers. After obtaining his degree, Antúnez de Mayolo toured Europe, going on a pilgrimage to industrial centers such as the Alioth Workshop in Switzerland, the Stassano steelworks in Italy, and the Ruhr Valley in Germany. He finished his tour in Norway, which perhaps had the greatest effect on him. "After visiting Oslo . . . we became convinced that we could do something similar in Peru, rich in waterfalls like Norway . . . to create a permanent source of wealth where there is none, through the magic wand of electricity."[14]

Antúnez de Mayolo returned to Lima, only to depart once again to take courses at Columbia University. Upon his definitive return to Peru in 1912, he struggled to find work in the public sector. At the new Ministry of Fomento and Public Works, he was told to travel to the Amazon and survey the rivers in that region to determine their navigability. Antúnez de Mayolo refused, as the salary did not cover living expenses and because of tropical diseases. That same year he carried out his own expedition to the Santa River, to study the feasibility of constructing a hydroelectric power station. Again, this proposal would be rejected by Fomento.[15] Continuing to focus on his studies, he received his doctorate in 1923 from San Marcos and briefly worked for Lima Light and Power (LL&P) before becoming a consultant for several mining companies. In the meantime, he also carried out groundbreaking theoretical work. He gave a lecture in 1924 titled "Hipótesis sobre la constitución de la materia," where he predicted the existence of the neutron, and in 1932 he established the existence of the positron—the counterpart of the electron—in his work *Los tres elementos constitutivos de la materia*. He made these discoveries before American and European scientists did but, despite being nominated for the Nobel Prize in Physics in 1943, Antúnez de Mayolo's contributions to the theoretical study of physics have largely been forgotten.[16]

While international recognition eluded him, he became one of Peru's best known scientists. It did not hurt Antúnez de Mayolo's career that his former classmate at San Marcos, fellow engineer Manuel Prado, reached the presidency in 1939. Prado's arrival marked the beginning of a close relationship between Antúnez de Mayolo and the Peruvian state, which would rely on his expertise on all matters regarding electrification. His proposal for the construction of a hydroelectric plant on the Santa River was championed by Prado, who made the project the cornerstone of his first administration. Antúnez de Mayolo was also called upon to survey other rivers to determine their hydroelectric potential, among them the Vilcanota River, future site of the Machu Picchu hydroelectric plant. In 1940 Manuel Prado directed the Ministry of Fomento and Public Works to study the Mantaro River, "so as to determine the hydroelectric possibilities that it offers, as well as the potential irrigation of the valleys in Jauja and Huancayo."[17]

Ráez y Gómez believed that the hydroelectric potential of the Mantaro was to be found in the Amazon, but Antúnez de Mayolo thought it was to be found in the province of Tayacaja itself. In November 1943, with his son Erick and a young engineer from the Santa Corporation named Jorge Bevin, Antúnez de Mayolo set off to explore the region.[18] Part of the province was *ceja de selva*, the borderland where the Andes ended and the Amazonian jungle began, and where violent changes in altitudes could be exploited. Since the Mantaro River surrounded the whole province, Tayacaja was referred to as a "peninsula." (See figure 17.) Antúnez de Mayolo began exploring the river as it flowed at high altitudes, until he came across a formidable curve and the river disappeared as it winded its way around a granite mountain. Antúnez de Mayolo then climbed the mountain to see the other side of the river as it violently descended into lower altitudes and into the Hacienda Villa Azul.[19] If the waters of the Mantaro could be diverted through the mountain, a great fall could be created, generating 1,000 MW of energy. As he climbed the Tayacaja highlands, the engineer came across the old imperial road that had been built by the Incas and used by the Conquistadors. This gave the enterprise an almost imperial feel. De Mayolo noted, "This road was followed by the Conquistadors on their march to Cuzco, always seeking gold, without suspecting that there was potential wealth in the river, greater than all the gold of the conquest, a wealth inexhaustible throughout the centuries."[20]

Examining the river, Antúnez de Mayolo pictured the future plant. At the river's curve, a reservoir would be built to regulate water flow, and a twenty-kilometer tunnel would be constructed to divert the waters. Once the waters reached the tunnel's end, they would plunge in a kilometer-long fall, powering the turbines on the other side of the mountain, before returning to the river. The difference in altitude allowed for the construction of two plants, one directly beneath the other, much like the *andenes* that surrounded him. The project was promising enough that the National Air Service surveyed the region, adjusting Antúnez de Mayolo's calculations by a mere two kilometers. Finishing his survey, he named the future station "Pongor," which was the

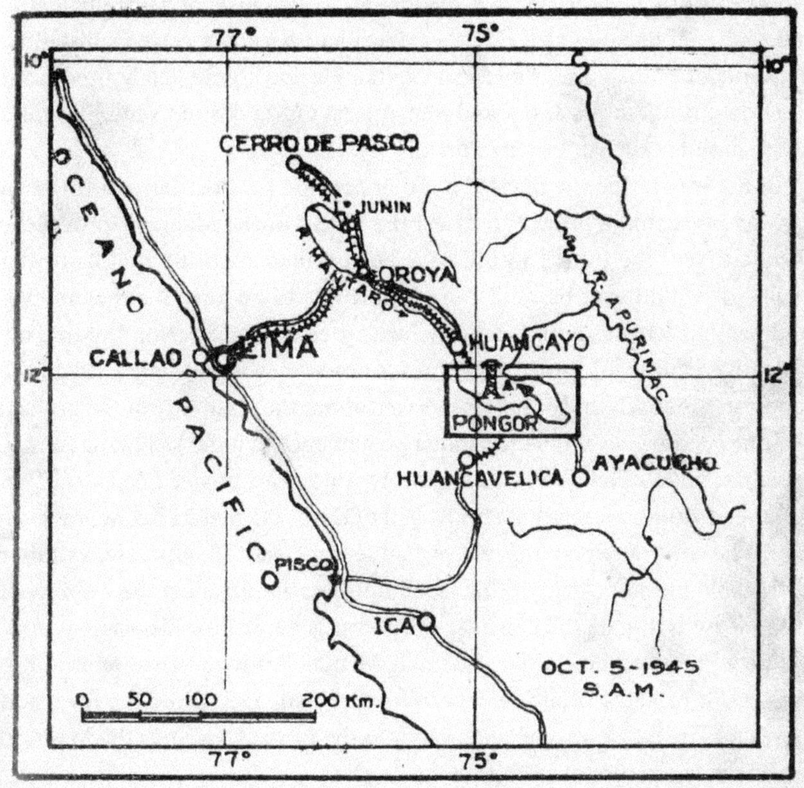

FIG. 17. Mantaro Plant zone of influence, according to Antúnez de Mayolo. Santiago Antúnez de Mayolo, *Memoria: Trazo preliminar de la Central Eléctrica del Pongor en el Mantaro* (Lima: Rímac, 1945). Pontifica Universidad Católica del Perú, Colección Especial.

name of the mountain where the plant would be built.[21] Coincidentally, it was also the name of a temple in Ancash dedicated to a power god venerated by pre-Inca civilizations.[22] It would not be surprising, given his interest in pre-Columbian civilizations, if Antúnez de Mayolo made sure to pay homage to the ancient deities of the Andes even in his scientific endeavors.

In the following years, Antúnez de Mayolo fell out of favor with the Peruvian state. When Manuel Odría arrived in 1948, he defunded the Santa Corporation, as he was skeptical of state-sponsored development projects in general, hoping instead to strengthen Peru's export-oriented model. Antúnez de Mayolo, having previously been involved in such development projects, wrote a scathing editorial in *El Comercio* accusing Odría of damaging Peru's international prestige, as the Santa project had already received substantial amounts of French and American capital. He was immediately fired from the board of directors and joked that at least Odría did not send him for a short "holiday" at San Lorenzo prison.[23]

In his frustration, he devised another scheme for the Mantaro River. In 1953 he presented a project to divert the waters of the Mantaro toward the Rimac River. The tunnel would have been unlike anything that had been built—or would ever be built—in Peru. It was to be forty kilometers long and pass under the central railway, Peru's greatest engineering feat to date. This diversion would solve all of Lima's power needs and help irrigate the lands on the southern coast. When asked about the feasibility of the project, Antúnez de Mayolo simply reminded his audience that the Delaware Tunnel, which supplied New York City with its water and power needs, was 136 kilometers long, noting that almost half of it was 137 meters below sea level, which he considered an impressive engineering feat.[24] In any case, Antúnez de Mayolo did not clarify if this second project would affect the viability of his first study, but he did consider the power generated by his original project to be "unmatched." As his classmate Manuel Prado returned to power in 1956, both projects would be vigorously debated, and, while the diversion never took place, it was debated well into the 1970s, even after the Mantaro hydroelectric project was built.

In time, Antúnez de Mayolo's proposal to build a hydroelectric plant on the Mantaro River would become one of the most divisive issues in Peruvian politics, a development that the "wise man" could not foresee when he wrote his memoirs in 1957. For the time being, it remained an imaginary of a great potential conquest. First the Incas had conquered the region and built their famous road system. The Spanish used those same roads to conquer the Inca capital with its riches. But now, science would conquer the Andes a final time, as the electric power generated by the river was the ultimate form of wealth, one that was renewable and could potentially transform the central Andean region into one of the great industrial centers of Peru.

AUTHORIZING A CORPORATION

After lying dormant for almost a decade, interest in the plan received renewed attention during the planning frenzy of the late 1950s and early 1960s. The first party interested in taking advantage of the waters of the Mantaro was the U.S.-owned Cerro de Pasco Corporation (CPC), when it contacted the

Ministry of Fomento and Public Works in 1954. Since the beginning of the twentieth century, the CPC had established its own hydroelectric plants to power its mining activities. By the end of the 1950s, the company wanted a more stable provision of energy and considered the Mantaro Plant to be a viable option. However, after carrying out a partial survey, the company considered the project to be too ambitious and quickly lost interest in it.

The French-crafted National Electrification Plan of 1957—to which Antúnez de Mayolo was a consultant—once again spurred interest in the project. The Mantaro scheme, according to the French engineers, would be the centerpiece of electrification endeavors in the central region, thus far dominated by Cerro's electric system. The plan recommended creating a single interconnected system that would benefit all miners in the area. The French experts believed that such a feat would be difficult because of the economic disparity between the CPC and small- and medium-sized miners but that, nevertheless, it would be best "from a national standpoint."[25] The project would also benefit the city of Huancayo, which would be directly supplied with electricity from the plant, putting an end to its constant power shortages. Finally, Lima—whose energy demand was constantly expanding—could also absorb a sizable amount of the electricity that would be generated by the waters of the Mantaro, although it was not meant to be its main beneficiary. This project would represent 42 percent of the electricity generated by all the plants encompassed in the plan, yet it required only 31 percent of the total investment. The French engineers, coming from a background in which the public sector played a key role in the development and management of infrastructure, suggested that, while it was convenient to grant a concession, the state should reserve some of the energy for its own developmentalist aspirations, as "no indemnity will compensate Peru [for] the opportunity of having a source of low-cost energy, which would favor its own progress."[26]

Soon after the plan's publication, several technical missions offered to further examine the hydroelectric potential of the river. The first was the German mission headed by Alfred Buch, a representative of the West German Bureau for Technical Assistance, who had been active in Brazil, India, and Venezuela.[27] The German study was followed by the Japanese Overseas Technical Assistance Mission, led by Tsuguo Nozaki. Nozaki, who had thirty years' experience in the electricity sector, considered the Mantaro project to be one of the most "ambitious hydroelectric projects in the world."[28] Both missions had the purpose of studying the technical feasibility of the project, but they did not offer any concrete development plans as to who would benefit from the power of the Mantaro. This gap was filled by a controversial study carried out by the Arthur D. Little consulting firm in 1960. Founded by the Bostonian Arthur Dehon Little in 1909, the firm focused on industrial research activities. Commissioned by Manuel Prado at the urging of his then minister of fomento, Alfonso Rizo Patrón, Little and Associates was to carry out a regional development plan for central and eastern Peru. The mission was headed by Murray D. Bryce, a leading figure on industrial development, and included other experts such as William Voorduin, president of the Regional Development Engineering Corporation of New York, in charge of studying the energy potential of the region. Equally essential was the work carried out by the Canadian firm Hunting Associates Limited, who undertook aerial photography and radar height profiles for the study.

The need for accelerated industrial development, argued Little and Associates, was urgent. While decades earlier there was concern over a manpower shortage, the "Little Report," as it became known, now stated that Peru faced a veritable demographic explosion. Peru's population would rise by three to four million in the 1960s and would double within twenty years. If industrial development was not fostered to provide better jobs, the report warned, Peru's political stability and economic survival would be "seriously jeopardized by the population explosion, the subsistence standards of living of more than half of the population, and the concentration of wealth in the hands of a few."[29] Criticizing Peru's suspicion of government intervention—the word "state" was avoided—the report advised that the country assume a leadership role and assist in promoting development. Yet Little and Associates stopped short of promoting direct government action, recommending an equally liberal approach in which the state should limit itself to preparing technical-economic feasibility studies and attracting local and foreign investors. Other than suggesting the creation of a development corporation—symbolizing the apogee of "mixed societies," in which the government would organize infrastructural projects but partner with mostly foreign capitalists to finance them—the state's role remained unclear.[30]

The report's main proposal was the creation of a development region known as "Peru-Via," a scheme to open a pathway for colonization toward the Amazon basin and, in the process, develop the central highland region, seen as a doorway to the east. Thus, despite being presented in modern developmentalist terms, Peru-Via reflected a dream that Peruvian statesmen and intellectuals had pursued since the end of the nineteenth century: to reach the "El Dorado" beyond the highlands.

The area encompassed by Peru-Via was chosen because it was close to the capital, as it was situated 150 kilometers east of Lima. This was considered essential as the Lima metropolitan area was the only viable market for what the region could produce. To the north, Cerro de Pasco and its mining activities and, to the south, the urban centers of Ayacucho and Cuzco would also benefit from the project. Interestingly, the main urban center within Peru-Via, the Jauja-Huancayo region, would be impacted by the development of light industry but was not to be a part of the plan, "although it is important to it as a route of access."[31]

The report struck a nerve with Lima's elites. The demographic explosion that the Little Report spoke of was already being felt in the capital, as new migrants settled on the outskirts of what was previously considered a modern and "European" city. Likewise, calls for agrarian reform, which had gained strength by the 1960s, would only intensify as the population of the Andes grew and the land question remained unsolved. "Peru-Via" could solve both these social and economic anxieties. The excess population could work new jobs in the region or could head east, thus stopping the westward internal migratory flows that had given rise to Lima's barriadas. Likewise, as migrants settled in the new lands in the Amazon, elites could avoid the need to radically redistribute land in the rest of the country, lending credence to the government's weak stance on land reform, although this subject was not touched upon by the report.

Regional elites had different motivations. Although it would be only a "route of access," Huancayo's elites supported the project because the report argued that the future of Peru-Via hinged on the construction of the Mantaro Hydroelectric Plant, which resonated with the city's industrial aspirations. The report stated that the "Mantaro River Power Project" was "probably Peru's greatest single known resource that is ready for early development."[32] It stated that the plant should be built regardless of the acceptance of the Peru-Via project, since the current demand for energy alone justified its construction and, if properly planned, "power might be available at approximately one-third of the cost per KW of TVA power."[33] While the project would promote agricultural and industrial development in the central highlands in the long run, it could be made economically viable in the short term by selling its power to regions in the west and to the coast, with LL&P, CPC, and the Marcona Mining Company listed as potential buyers. *La Voz de Huancayo* celebrated the report's findings and urged the government to adopt its recommendations. Given the decentralized nature of the project, the newspaper stated that Peru-Via "will surely count [on] the support of the whole nation . . . because for the first time a project is structured that gives real impulse to the progress of the Andean regions."[34] Furthermore, Huancayo's elites anticipated that the construction of the plant would reinforce the city's advantageous central geographical position, placing it at the forefront of Peru's economic development.

The debate also took place in the context of the arrival of the Alliance for Progress and its academic basis, Modernization Theory. Walt Rostow believed that economic development weakened communist tendencies, and this development was to be fostered by infrastructure, which helped a country move through different stages of growth.[35] Such ideas were also used by provincial elites writing for *La Voz de Huancayo*, who thought that the Mantaro Plant would serve as a bulwark against communism and be more effective than the anti-communist laws being passed in Lima. "The only way to fight communism is to raise the standard of living . . . for the communist fire is fed by the dry straw of discontent and misery."[36] But when exported to Peru, Modernization Theory interacted with elements of Dependency Theory, which would reach its apogee in the 1960s and 1970s. Dependency Theorists argued that underdevelopment was not only the result of local conditions but the direct result of the development of first-world countries. In this sense, at least for the elites writing for the city's newspaper, development was seen as a way of not only fighting communism but also resisting economic dominance from the first-world, for industrialization meant "the economic liberation from manufacturing nations and the overcoming of underdevelopment and colonialism imposed by the political economy of the strong."[37] Peru's quest for development, hence, was not only necessary to fight communism but was also an end unto itself.

The project was met with excitement in the national press, but it remained unclear if it would find acceptance in Peru's fragmented political scene. In a strange twist, it fell to Pedro Beltrán to present the project to the Chamber of Deputies. Beltrán was the leading figure among Peru's agrarian elites and had long advocated against state-led initiatives for development given his background as a staunch economic liberal. When Manuel Prado began his second term, Beltrán, through his newspaper *La Prensa*, had become one of the government's fiercest critics. In a shrewd political move, Prado invited him to become minister of finance and head of the Council of Ministers. Beltrán was able to halt the deterioration of the country's economy through orthodox measures, but he now had to justify the construction of Peru's largest infrastructural project. At first, he offered rational economic justifications, stating that while at first hydroelectric energy requires greater investment, "it can continue to function without exhausting its resources and using a force that does not cost or consume: the water from the heights."[38] Likewise, the use of hydroelectricity meant that Peru would no longer depend on oil, an explosive issue because of the relationship between the Peruvian state and the U.S.-owned International Petroleum Company (IPC), which, according to nationalists, was robbing Peruvians of their oil.

Beltrán also adopted Modernization Theory language, presenting Mantaro as nothing short of a "revolution" and a "radical transformation." Indeed, together with Peru-Via, it represented a new beginning, which would lead to industrial growth, thanks to its cheap electric power. It is also possible that Beltrán, one of Peru's most notable landowners, hoped that this would make land reform irrelevant—a process which he blocked time and time again as head of the Commission for Agrarian Reform and Housing—for Peru-Via would open new lands, "where a variety of climates would result in a diversification of agricultural production."[39] Yet, Modernization Theory also argued that "revolutionary" changes in agriculture were needed for economic "take- off," which would have meant reforming some of the most archaic aspects of land tenure in Peru. In that sense, Peruvian elites took one of Rostow's preconditions—investment in infrastructure in the form of Peru-Via and Mantaro—to avoid the other: land reform.[40]

The liberal Beltrán's "revolutionary" depiction of the Mantaro project was met with skepticism, particularly by the Christian Democrats (DC) and AP, two parties that sought to reform Peru's stratified economic and social structures. Once again, the most outspoken critic of the project was DC congressman Héctor Cornejo Chávez. Skeptical of Beltrán's pitch, Cornejo Chávez mocked notions of the Mantaro Plant contributing to an economic "take-off." He went as far as ridiculing congressman Eduardo Watson Cisneros—head of the government budget committee—telling him that "the 'take-off' . . . should be akin to an airplane, but it is looking like the efforts of a duck which will not take flight." This was met with laughs and cheers in the chamber. Clearly offended, Watson responded, the term "'take-off' to which you have so humorously referred, is not an idea of mine, but the idea of a well known economist named Rostow." In turn, Cornejo Chávez congratulated him "for not coining the term."[41]

Beltrán's presentation highlighted the differences among Peru's political class, but it would be up to the minister of fomento to secure final approval of the project. By then the new minister was engineer Jorge Grieve, a member of the American Popular Revolutionary Alliance (APRA). APRA had originally been founded as a revolutionary and anti-imperialist party, but by the late 1950s it had sought a policy of moderation and even cooperated with the oligarchic Manuel Prado during his second term, a period known as "La Convivencia." While Grieve was not as unpopular as Beltrán, his appearance in Congress proved to be equally controversial. Independents and members of Prado's Movimiento Democrático Pradista (MDP) supported Grieve, while the AP-DC bloc sought to approve the project on their own terms. Stressing the importance of regional development, Grieve stated that, thus far, the consumption of electricity was a dynamic that had only benefited cities and a certain "political class." Indeed, in the 1940s, Lima, with a population of half a million, could boast a figure of 125 watts per inhabitant, while the rest of the country, with a population of six million, enjoyed a mere 6 watts. By 1960, Lima, despite growing threefold, had maintained the figure of 120 watts, but the rest of the country—now with a population of over eleven million—had only risen to 17.5 watts.[42] He also defended Prado from accusations that the Mantaro project was essentially a political enterprise, destined to ensure Prado's legacy, and denied that it lacked proper planning. In fact, he stated that the electric sector was the only sector in which an actual plan existed.

After a presentation filled with graphs and figures, Grieve concluded by saying that the sad situation in the Peruvian highlands was caused by a lack of mechanical energy, as the energy produced by animals or humans had, using Rostow's words, forced Peru to remain at an early stage of development. But when speaking of the challenges faced by Peru, the minister inevitably turned to examples such as China and, especially, Russia, countries that had not followed Rostow's path. Grieve would conclude by saying, "As Lenin stated in 1920: Communism is Soviet power plus the electrification of the country."[43] Despite the government's anti-communism and Modernization Theory parlance, it appears Lenin's prophecy still dominated the imaginaries of the proponents of electric development, especially for an old *aprista*.

A heated debate took place before and after the presentation, one that shone a light on both new foreign ideas of development and old, established domestic perceptions of fomento, or promotion. The opening salvo by the opposition questioned the benefits of the project in economic terms. Would it indeed be a hydroelectric plant that would spur (*fomentar*) development, or would its power be consumed by existing industrial activities? Grieve stated that the first phase of the plant would benefit existing activities in the area, such as CPC, Marcona, and LL&P. If that was the case, as opposition members pointed out, the Mantaro project was destined to subsidize dominant foreign groups rather than promote "true" Peruvian development.

Others openly supported the idea of fomento itself, such as Carlos Bisso Loredo, an independent congressman from Lima. Bisso Loredo had traveled to China as part of a diplomatic mission and stated that during his visit to the province of Fujian, he saw the Chinese construct a plant that would produce 800,000 KW, "without knowing how to industrialize that energy output." Bisso Loredo also mentioned the Russians: "in all the sites that the Volga allows, where there is a fall of 20 meters, they immediately build a plant, and it does not mean that there is industry ready to absorb the energy."[44] Others, not limited to members of Prado's MDP, pointed out that in Iran the government had moved fast to build the Dez Damn—Iran's largest hydroelectric project before the 1979 revolution. They did not wait until development studies were finished before approving its construction, as it was agreed that the project would be beneficial despite the lack of an integral plan.[45] Following a long Peruvian infrastructural tradition, which was also informed by a broad knowledge of international development projects, many simply thought that if electricity was available, then industry would somehow flourish.

Unsurprisingly, the debate became heated when the administration asked for executive freedom to seek financial support for the project. Opposition congressmen feared that the government would favor some of Prado's European contacts and pointed out that construction should be financed by an international organization that offered favorable conditions. It once again became clear that the real crux of the debate was not the project itself, but rather how much freedom should be given to the executive to pursue it. Given that Prado's term was set to end the following year, most members of the chamber were unwilling to support such an ambitious project for an outgoing government. The conclusion was reached that an autonomous entity should be created to oversee the project. Grieve and the government did not oppose the initiative; indeed, they supported it, as it would ensure the project's continuity after Prado's departure.

But even on this point there was friction, as some wanted to call the new entity an "authority" and avoid the name "corporation." This was not simply a semantic argument, as Peru had had an almost traumatic experience with public corporations destined to spur industrial development. The most well known case was the Peruvian Santa Corporation, established by Prado during his first government and meant to establish a steel industry in northern Peru with the aid of electrification. The Santa Corporation always struggled to "take-off," and by the 1960s most of Peru's political class considered it to have been a failure. Indeed, when the National University of Engineering held a forum some years later to discuss the merits of the Mantaro project, the dean of the university—Mario Samamé—stated in the forum's opening remarks that it was necessary to avoid the "sad experience" of the Santa Corporation.[46] Despite the failure of the agonizing Santa Corporation—the point of reference for the establishment of any similar experiment in the future—DC congressman Jaime Rey de Castro López de Romaña would reject the idea of naming the organization an "authority," considering it "a vulgar copy of the authorities that exist in the United States. Our traditional name for this type of institutions—despite acquiring a bad reputation—is 'Corporation.' It is not necessary to go to other countries to import names. We could also copy the Italians and call it the Electric Front of the Mantaro. 'Corporation' reflects our psychology, our traditions, and our customs."[47]

The creation of the Corporación de Energía Eléctrica del Mantaro (CORMAN) at the end of 1961 was greeted with enthusiasm, but many remained skeptical. It became clear that the government had used Peru-Via to justify the Mantaro project. It dismissed the idea of establishing a development corporation, instead creating the Interministerial Commission for Development—headed by Rizo Patrón, the main champion of Peru-Via—which quickly faded away. More importantly, the press criticized Prado for not reaching an agreement with the World Bank's International Bank for Reconstruction and Development (IBRD), although the latter refused to finance the project without further studies as it considered it too ambitious for Peru's current level of development as well as lacking an integral plan.[48] Likewise, as Peru-Via required a separate source of funding, it became disconnected from the Mantaro scheme and was eventually discarded by the government.[49] It quickly came to light that the government had reached an agreement with European capitalists to finance the project, as it had struck a deal with The English Electric Company, George Wimpey and Co. Limited, and the German firm Siemens-Schuckertwerke.

The deal with European capitalists had likely been made when Prado visited West Germany and Great Britain the previous year. Prado's visit to Europe was an essential part of his "Prado Doctrine," a foreign policy that sought to move away from the United States and push Peru closer to Europe.[50] Unlike his first presidency, in which collaboration with the United States had elevated Peru's regional standing, Prado could show few concrete accomplishments in his new pro-European foreign policy except for the technical missions and contacts that he had made during his trip. Thus, Prado's foreign policy found real expression only in infrastructural development and securing financial backing for the Mantaro project.

Given the estimated cost of the project—ninety million U.S. dollars, an unprecedented sum in Peru's history—the media questioned the transparency of the contracts. There was no final feasibility report on the Mantaro power project, and whatever estimates had been done in the past would inevitably change given the long-term nature of the enterprise. Nor was it clear if the financing of the plant would consider the whole development of the region. *El Comercio*—which in the 1960s had a nationalist element in the figure of Alfonso Miró Quesada, a rogue figure in the otherwise liberal family that owned the newspaper—stated that it was "curious" that the government had rushed to sign with the Europeans.[51]

Others were exceedingly generous regarding Prado's accomplishments. Peruvian journalist and Beltrán ally Jorge Luis Recavarren stated that the Mantaro Plant was a policy of "healthy and efficient patriotism," precisely because it was destined to serve the future needs of the country and not result in immediate political benefits. It was, above all, nothing short of an authentic "revolutionary" undertaking, for "the best revolutions are those that are carried out peacefully, with the aim of securing the progress of the collective," and, thanks to this new concept of revolution, the country could cast aside "nineteenth century notions of bloody and chaotic revolutions."[52] Mantaro thus seemed to represent what Michael Latham has argued was the sort of change that Modernization Theory sought in the context of Cold War politics, which was "a right kind of revolution."[53]

Beneath the revolutionary and modernizing tone, however, it soon became clear that for Prado this revolution was meant to avoid more serious reforms and, indeed, maintain existing social, economic, and political structures. Shortly before leaving office, Prado celebrated that the Mantaro Plant was to shift the Peruvian economy from an agrarian to an industrial one, rendering agrarian reform irrelevant. While his own family interests were mostly financial and industrial, he did not wish to antagonize landowning elites. He did, however, create a land reform commission, which, as previously mentioned, purposely achieved next to nothing. Furthermore, while the provinces would be developed, Lima was to remain the political, economic, and social center of Peru. "The Mantaro river's extraordinary flow will not only satisfy the industrial needs of Junín, Huancavelica and Ayacucho, but, in addition, the needs of Greater Lima in the coming years, when it becomes the great industrial center that it is meant to be, maintaining its hierarchy as capital."[54] Whatever decentralization was to take place was purely economic, not political. And even in this sense, the economic benefits *must* also benefit the capital, reinforcing the status that it had held since the viceroyalty.

Mantaro was to not only maintain the political preeminence of Lima but also uphold traditional social and racial hierarchies *within* the city. Prado would again echo this sentiment when financial backing for the project had been secured, ironically as he was inaugurating public lighting in the new districts on the outskirts of Lima. "With cheap electric energy it will be possible to achieve the second great exodus of Peruvians, which will be the return to the native lands of those who had journeyed to Lima and other great cities."[55] If half a century ago oligarchs feared that electricity would destroy Lima's mystique, now it was seen as a useful tool for preserving a city that was becoming unrecognizable to its elite. Manuel Prado spoke of development and modernity, no doubt promoting them with sincerity, but in many ways he wished for a "revolution" that would reinforce the status quo. Little did he know that he was fighting a losing battle, as migration from the countryside to the city would only intensify, and no industrial process—even if successful—could occur quickly enough to stop the Andean exodus.

THE RETURN OF THE ITALIANS

Prado rushed to secure financial backing for the project that was to be his legacy, but the armed forces deposed him a few weeks before the end of his term. The intervention of the military had little to do with Prado's actions. It happened because no clear winner had emerged in the 1962 elections, in which APRA—an old enemy of the armed forces and now allied with conservative oligarchs—almost took power. While the junta announced that it would hold power for one year before handing it back to civilians, it was by no means a caretaker government. The military enacted a pilot land reform program and created the Instituto Nacional de Planificación (INP), hoping that such tools would be used by a future civilian administration to carry out substantial reforms and thus avoid the possibility of a communist revolution.

Despite the political rupture, CORMAN's work continued. The entity was not absorbed by the INP and was only briefly mentioned in Peru's first social and economic development plan, released in late 1962, where it was stated that the power of the Mantaro was simply to foster the central region's agricultural and industrial development and to substitute oil motors in mining.[56] As an autonomous entity, CORMAN did receive funds to carry out preliminary work on the site, such as roads and exploratory tunnels. Equally importantly, its board of directors was finally appointed. It brought together notable technical personnel from both the private and public sectors, giving the corporation a technocratic feel. Alfonso Montero Muelle, a successful businessman and the former director of the National Society for Industry (SNI), was named head of the board. From the public sector, representatives from the ministries of Fomento, Agriculture, and Hacienda were included. From civil society the SNI and the National Society of Mining and Oil were present, as well as one representative from the various existing fomento banks. Finally, the Peruvian Electrotechnical Association (AEP) and the Association of Electrical Businessmen (AEE) sent one member each. The armed forces, eager to be represented, named a young colonel, Francisco Morales Bermúdez, as its delegate.[57] Junín's elites, much to their dismay, found themselves completely excluded. Given that the Mantaro operational offices were established in Huancayo, the city's political and economic elite—particularly those belonging to the Huancayo Chamber of Commerce—furiously complained about their exclusion and wrote several letters to the government, which went unanswered.[58] While the project was presented as one that would develop the central highland region, it was clear that it would be directed from Lima.

Perhaps the most crucial development during the short-lived military junta was the approval of the definitive technical study of the project, which was to be carried out by the firm Electroconsult of Milan. Submitted to the Peruvian government in early 1963, the Italian study estimated that the cost of the Mantaro project was considerably higher than anticipated. The new estimate, 196 million dollars, was 100 million more than the figure that had been presented by the previous government. This meant that the contracts with the Anglo-German consortium would have to be revised, or potentially annulled.

In the meantime, the plant's construction advanced slowly, but support and opposition for the project quickly intensified. Antúnez de Mayolo became involved in the debate when he wrote an editorial in which he stated that the Mantaro project would not only offer Huancayo the opportunity of developing its industries but would also turn the region into a tourist center. Thinking as an engineer, Antúnez de Mayolo anticipated that the colossal project would attract visitors and floated the idea of turning the rustic buildings of the former Hacienda Villa Azul into a tourist hotel. Since roads in the region would have to be expanded and improved to move heavy machinery, they could also be used to reach the site without difficulty. The hotel would be fitted with a casino, beautifully designed gardens, and tennis courts. At night, a light show would illuminate the waterfall and the hydroelectric plant. "During the day, the whole system would appear like a chiffon dress in which the mountain will be the bride, and, during the night, the pumps and the waterfalls would provide a fantastic effect, a unique spectacle in [the] world." Antúnez de Mayolo even suggested that the Peruvian state advertise the plant in television commercials to be broadcasted in Europe and the United States. This infrastructural tourism would benefit Huancayo, to which the "wise man" extended his gratitude for the "warm welcome extended by the city . . . because of our initiative of the Pongor project."[59]

La Voz de Huancayo continued its campaign in support of the project, dedicating weekly editorials to the cause of the hydroelectric plant. As it lashed out against those who opposed the project, an idea that a conspiracy was afoot began to emerge. "There is no shortage of people interested in creating difficulties for the realization of this enormous project which the future of the country hinges upon."[60] The newspaper stated that these enemies had always existed and even commented that when Antúnez de Mayolo first reconnoitered the site, sensing the magnitude of the project, he must have felt like "Don Quixote." Twenty years after Antúnez de Mayolo's expedition, the sense of impossibility and fatality remained.

The people of Huancayo were not unjustified in their fears. At least two actors were dubious as to the merits of the project. According to American officials, Carlos Mariotti, manager of LL&P, feared that the company might be forced to buy energy from Mantaro and might even be absorbed by the new network. Having enjoyed almost complete autonomy since its inception and accustomed to expanding at its own pace, LL&P did not take kindly to the state becoming a player in the electric field and criticized the newly created bureaucracy of CORMAN, stating that CORMAN had a vested interest in pushing the project through. The company quickly presented the government with its own plan to build a plant in the Huaura River north of Lima. The plant would produce 300,000 KW—equal to the first phase of Mantaro—and reclaim through irrigation 15,000 hectares of land. If the project did not materialize, the company might indeed have to "purchase 70% of Mantaro('s) output (by) 1969." Unsure of whether the Mantaro project would go through—a valid concern given the many projects approved but not finished by the government—LL&P would take a "wait and see" attitude, a difficult position for a company that had a carefully crafted expansion plan. However, fearing a backlash from the public, the company "was not in (a) position to oppose . . . (the) Mantaro project."[61] The synergy that had characterized the relationship between the company and the state was disappearing.

The second actor who opposed the project was the CPC. It informed the State Department that it was "seriously concerned" about the actions of the Peruvian government, as it thought that it had not assessed the project "intelligently." Other specific objections included Cerro's concern that if the waters from Lake Junín were diverted toward the Mantaro project the company would not have enough water during dry seasons for its own plants. Indeed, the company expected that the government would eventually force Cerro to purchase power from Mantaro. Finally, the company concluded that the Peruvian government had shown a "complete disregard for Cerro's interests and acquired rights."[62]

The return of democracy, with the arrival of Fernando Belaúnde Terry and his progressive AP party in 1963, also posed some challenges. Belaúnde had promised a "revolution without bullets" in his quest to reform Peru's rigid economic and social order. However, Belaúnde's revolution centered around "traditional" infrastructural projects such as road building and irrigation, especially the Majes and Olmos River projects. Likewise, Belaúnde spoke of a "revolution of credit" that would facilitate Peru's full transition to a capitalist economy.[63] Despite many AP members supporting the Mantaro scheme—as it strongly resonated with his idea of the "Conquest of Peru by Peruvians"—Belaúnde had other priorities. Indeed, this became clear during a visit by Walt Rostow to Peru in 1966 to improve relations between the two countries over the IPC dispute. President Belaúnde had taken the economist to visit his new highway construction project in the Amazon, and not the site where Mantaro was to emerge.[64]

Although skeptical, the new president did not wish to wholly abandon previous projects, as he feared that he would be accused of dismissing them simply because they had been carried over from the previous administration. The stage was thus set for one of the most contested political battles of the Belaúnde administration, although it would later be overshadowed by the IPC negotiations and the failure to achieve significant land reform. In Congress, a coalition of apristas and supporters of former dictator Manuel Odría fought bitterly with Belaúnde over every issue, and the latter's reluctance to back the Mantaro project was used to great political effect by his opponents. It fell upon Ramiro Prialé, a Huancayo native and a leading aprista, to challenge the government as president of the Senate. Prialé stated that there was a conspiracy against the project by foreign interests, represented by Mobile Oil, which was interested in exploiting the newly discovered oil fields of Aguatía.[65] Once again, foreign oil companies were threatening Peru's development. There were also local forces at work scheming against Mantaro, who—according to Prialé—had managed to gather $60 million to stop the plant's construction. The senator pointed a finger at LL&P and at Juan Carosio, apparently unaware that Carosio had passed away in 1959.[66]

Support for the project in Huancayo acquired greater urgency as the "red menace" that emerged after the Cuban Revolution seemed to become real when Luis de la Puente Uceda and the Revolutionary Left Movement (MIR) began their activities in the jungles of Junín. Although there is no evidence that the guerrillas attacked the construction site, their bases were close to Tayacaja. Given that the region was plagued by "misery and unemployment," the construction of the plant would ensure that the guerrillas could not count on local support. However, Huancayo's elites thought that the project had economic merits of its own and would have supported it just as eagerly even if no communist threat had existed in the area.[67] Soon enough, under the leadership of the former mayor of the city, Luis Caballero Alvarado, Huancayo formed a "Defense Committee of the Mantaro," whose purpose was to pressure Junín's congressmen to take a tougher stance in Lima. It urged their representatives to follow the example of Ramiro Prialé. The timing of its formation was crucial, as the government had until the end of January of 1966 to decide whether the Anglo-German consortium would carry out the project's construction.[68]

Sensing an opportunity amid the political chaos, an Italian group, GIE-Impregilo, presented a new bid that was more attractive than that presented by the Anglo-Germans. Faced with political pressure from all sides, Belaúnde decided to rescind the existing contract and go with the Italians. After all, GIE-Impregilo had solid credentials, having successfully carried out ambitious hydraulic works in Africa and Asia. The signing of the new contract was met with relief by supporters of the project, but with considerable dismay by the British and Germans. A diplomatic crisis ensued as the British Embassy issued a strong rebuke and told the Peruvian government to expect less aid in the future, as the annulment was even covered by the British press.[69] As for the Germans, they threatened to withdraw their ambassador from the country.

The diplomatic fallout was considerable enough that *The Economist* published an article on the dispute that was reprinted in the local press. Stating that "Latin American politics and a Latin ingenuity for business took advantage of the British and Germans," the influential publication commented that Belaúnde had not been keen on the plant, as it would interfere with his cherished irrigation projects. It also pointed out that the Anglo-Germans—blindsided by both the "Latin" Peruvians and the Italians—had found out about the cancellation of the contract from the press.[70] It also shone a light on the cutthroat competition in the business of international development, as Electroconsult, the Italian firm that had carried out the feasibility studies, had links with Fiat, who in turn had links with GIE-Impregilo. This was not the first time that the Italians had outsmarted their competitors, as they had also won the bidding on the Machu Picchu hydroelectric plant through the Italian firm Panedile in 1958. But Belaúnde's approval had its price. The Mantaro project would move ahead only if the president obtained the blessing of a hostile Congress for his other infrastructural projects.[71]

With financial backing in place, construction of the project resumed. It soon became clear that the plant's construction would be as difficult as the politics behind it. Tragedy first struck in 1964 when a young engineer named Zenobio Rodríguez Vivas fell off a 300-meter ravine, trying to escape a landslide caused by a controlled explosion gone wrong.[72] A year later, six men died when their vehicle fell off a cliff. The crash site proved to be so inaccessible that the bodies were not retrieved until forty-eight hours later. They were eventually buried in Lima, Jauja, and Pampas, the latter being where most of the workers came from.[73] In 1965 Juan Campos Montañez and two others died while working on one of the penetration roads. Two years later his mother, widow Isabel Montañez, sued CORMAN, as her son "was the only one who had taken care of her."[74]

The deaths of their fellow laborers pushed CORMAN workers to form their own union to fight for better conditions. The workers received support from Congress, and one of Belaúnde's own congressmen, Alfonso Mendoza Gálvez of Huancavelica, stated that Bertolero and Co.—the Peruvian company in charge of civil works—"was wasting government money" and mistreating workers, who had no access to running water or decent housing.[75] Mendoza Gálvez's denunciations had considerable ramifications, as representatives of Tayacaja at one point demanded to leave Huancavelica and be annexed to Junín. The construction of the plant created friction between the inhabitants and the local government, because Pampas—capital of Tayacaja and the main location of the construction site—did not have access to the most basic infrastructure. It did not have any roads that connected it with the capital, Huancavelica, but it was connected to Huancayo. Despite the Huancayo press depicting Pampas as the "Benjamin" of the department, they did not pursue the matter further as they did not want to foster a separatist movement.[76]

Soon, the population of Pampas multiplied as working camps were established. The first camp was Kichuas, built near the future site of the reservoir, the other being the larger Campo Armiño, located on the other side of the "peninsula," where the machine houses were being erected. Although the number of people attached to the project never exceeded 2,500, the population of both camps at one point reached 10,000 souls, as nearby inhabitants flocked to the site for employment opportunities. The new Italian administrators began to make sure that workers lived in decent conditions, and preassembled houses were quickly erected. These population centers enjoyed all the amenities that the people of Huancavelica did not have access to, such as a hospital, a school, *comedores*, bakeries, a cinema, and a soccer field.[77] As for the Italian administrators, being in the midst of the Peruvian highlands of the Tayacaja "peninsula" did not impede them from living as they did in their own peninsula, as they imported pasta and wines from Italy.[78]

The camp also became a fertile site to carry out social experiments, as an educator from the Pontifical Catholic University of Peru arrived to foster greater "harmony" for workers' families as part of a dissertation on family education. Arriving in Campo Armiño, the educator found high levels of alcoholism, despite lower numbers for cigarette smoking and use of coca.[79] Presented with these challenges the educator wrote about empowerment and maintaining some social cohesion in the camp for families. She recommended that workers' wives learn trades such as knitting and embroidery, flower arrangements, and other "small industries." As for the children, the importance of health, hygiene, and nutrition was emphasized. However, while science played a key role in her design, so did religion, for communion and marriage were key to a harmonious social order. She preached that workers should be married until their deaths, as "man cannot separate that which God had united."[80] The tensions between modernity and tradition were present in all the camp's activities.

Soon, heavy machinery began to make its way through the central Andes. Locomotives, tractors, perforators, and cranes were sent from Europe and arrived at the port of Callao. From there they would be transported to Pampas via the central railway. Once they arrived in Huancayo, they would be taken by large trucks to the construction site on rudimentary roads built by CORMAN. These trucks appeared to be delicately balancing on the cliffs so as not to plummet. This sophisticated machinery made the project possible, but difficulties abounded. In November 1967 waters rose in the tunnels when a provisional dam flooded, and a considerable amount of machinery and equipment was lost. Sometime later the CORMAN warehouse was pillaged, and 100,000 soles of dynamite were stolen. The guard in charge and a handful of workers were captured as police arrived on the scene.[81]

Tragedy struck once more when gas was discovered in one of the tunnels, claiming the lives of fifteen workers. This led to the establishment of a commission headed by Huancavelica congressman Víctor Freundt Rosell, and accident prevention courses were taught to workers. It also became apparent that relations between the Italians and the workers were not bereft of tension, as the CORMAN union threatened to go on strike if two workers from GIO-Impregilo were not fired. Mantaro began to claim the lives of Italian engineers as well. In June of 1967 engineer Davide Olivato drowned while carrying out an inspection of the tunnels.[82] Some months later, four more Italians lost their lives when their vehicle plummeted in Tayacaja's highlands. Engineers Olimpio Codazzicotasi and Adesio Badsi died, as did Codazzicotasi's son, nine-year-old Osvaldo, and a baby named Paulo Rosou. Making sure that nothing was left to chance, aside from offering safety seminars, the Italians erected the camp's first chapel.[83]

Other deaths marked the first phase of construction when both Manuel Prado and Santiago Antúnez de Mayolo died in 1967. By supreme decree it was stated that the plant would bear the name not of its political architect, but of its engineer. This was perhaps the only point that Congress and the executive agreed upon. But by the mid-1960s Belaúnde had to deal not only with a hostile Congress but also with elements of civil society that were attacking the government's mild developmentalism. The proto-neoliberal Centro de Documentación Económico-Social denounced the project as "too big to be stopped and, at the same time, too big and uneconomic to continue."[84] The group also criticized regional development plans, stating that, if done with care, they could be beneficial but if done for political reasons, could lead to the creation of inflated bureaucracies, CORMAN being an example. Considering that the costs outweighed the benefits, the Centro de Documentación Económico-Social went against the grain by stating that "electricity by itself does not create wealth."[85] While some may find the beginning of Peruvian neoliberalism in the 1980s, some of its forebears were already present in the 1960s, partly created by ambitious schemes such as Mantaro.

As economic difficulties mounted, Belaúnde decided to reduce CORMAN's funding toward the end of his presidency. The idea of turning CORMAN over to private interests, in which LL&P and CPC would have greater influence, was also considered. This caused a nationalist outcry in the press, and Carlos Mariotti simply stated that he had had informal discussions with the government, and that it was not the private sector's desire to take control of the project but simply to aid in its construction.[86] All the while, the workers at Pampas diligently continued their work. As they blasted through the rocks in the quest to finish the 20-kilometer-long tunnel, another event occurred that halted the progress of the Mantaro Plant, when on October 3, 1968, Belaúnde was deposed by the military. Unlike the liberal Prado, or the reformist Belaúnde, this new military government promised nothing short of a "real" revolution. The political meaning of the hydroelectric plant was to radically change in its last phase of construction.

REVOLUCIÓN CON ELECTRIFICACIÓN

The self-proclaimed "revolutionary" military government stated that it would radically modify Peruvian economic and social life through a "revolution from above." Equally rejecting classical capitalism and Marxist planning, the new regime sought to develop a unique homegrown model of development. A thorough land reform program was enacted, key sectors of the economy nationalized, Import Substitution Industrialization (ISI) policies promoted, and diplomatic relations with the Eastern Bloc established.[87] Although much ink has been spilled regarding these events, few have analyzed the government's energy policy, which in many ways reflected the "revolutionary" spirit of the military.

The GRFA lost no time in transforming the structure of the Peruvian state, as it sought greater action when it came to "planning, exploiting natural resources, basic industry, and essential public services."[88] Whereas previous liberal governments debated the place of electricity within the Ministry of Fomento and Public Works—whether it should be associated with industry or have a directorate of its own—the new government decided that Fomento itself had run its course. All previous responsibilities of the defunct ministry—road building, housing, and industry, among others—were assumed by new ministries. As for electricity, its place was radically altered in the bureaucratic structure in December 1968 with the creation of the Ministry of Energy and Mines.

The association of energy and mining was an odd choice for a government pursuing an ISI policy, and, ironically, it was reminiscent of the days of the Aristocratic Republic, when hydroelectricity was linked to the Directorate of Mines and Oil within Fomento. But the linkage responded to political priorities. Energy meant not only hydraulic sources of power but also oil. Given that armed forces carried out the coup under the pretext of ending the poisonous relationship between the Peruvian state and the U.S.-owned IPC, the new ministry wanted to emphasize the national ownership of Peru's subsoil minerals.[89] This also explains why the new minister was General Fernández Maldonado, a member of the radical faction of the government.[90] Hydroelectricity, thus, became associated with oil and mining, although it had little in common with either. Despite the bureaucratic shake-up, former functionaries of the Directorate of Industry and Electricity were absorbed by the new ministry, and bureaucrats who had begun their careers in Fomento decades earlier continued to work for the new government, who needed their technical expertise.[91]

In time the government led by General Velasco Alvarado would portray hydroelectric generation as one of the main pillars of the revolution, together with agrarian reform and the recovery of Peru's national resources, but the government could not take control over the sector until a steady supply of power was available to be directly provisioned to Peru's citizens and industry. This meant that, for the time being, private companies such as LL&P and CPC would continue to control their respective electric grids. It also meant that the military had a vested interest in speeding up the construction of Mantaro. The composition of CORMAN was modified, and a greater number of generals were incorporated into its board of directors. General Pedro Puente Revilla, a military engineer, became head of the corporation, which was quickly purged of all civilian elements—except for those representing the AEP—and reduced to eight members representing the three branches of the military.[92]

Ceasing to be a political battleground, construction of the plant accelerated. One year after taking power, the GRFA inaugurated the dam located at Kichuas. The dam's construction had presented severe challenges, as workers and engineers had to wait until the water was at its lowest then carry out a controlled explosion to momentarily block the river. Workers fought against time pouring concrete into the dam as the waters rose behind them. Named Tablachaca—a Spanish-Quechua amalgam that roughly translates as "bridge of planks"—the dam was one kilometer deep and two kilometers long, consisting of 157,000 cubic meters of concrete, situated almost 3,000 meters above sea level. To make sure that *huaicos* did not damage the structure, steel rods were inserted into the surrounding mountains. Two years later, the 20-kilometer-long tunnel, which had claimed the lives of more than a dozen workers, was inaugurated. For the occasion, General Fernández Maldonado, together with the Italian ambassador and the president of GIE-Impregilo, traveled to Pampas to celebrate its completion. A new face was also in attendance, as Soviet ambassador Yury Lebedev was invited to the ceremony.[93]

Yury Lebedev's presence symbolized the military government's closer relationship with the Eastern Bloc. Just as Manuel Prado had used electric infrastructure to advance his "Prado Doctrine," the GRFA used Mantaro to show the eager Soviets the advantages of economic and technical cooperation with Peru, which "became the center of Soviet operations on the continent."[94] Not long afterward, by late 1972, Peru and the Soviet Union signed a protocol in which the latter pledged assistance to aid Peru in the construction of the Olmos Irrigation and Hydroelectric project. The Soviets sent specialists and machinery and carried out studies, much to the concern of the State Department.[95] The Olmos project was eventually abandoned by the government, but it was an indication of how the GRFA, like its predecessors, was willing to use infrastructural development to strengthen its foreign policy goals.

As the project moved forward, so did the government's plans to exert greater control over the energy sector. As workers celebrated the culmination of the tunnel, the military published a global development plan, which included a new "revolutionary" conceptualization of electricity. The new discourse had been influenced by various intellectual trends that had been present at the Centro de Altos Estudios Militares (CAEM), a military training school established in 1950 with the goal of professionalizing the army. The most important idea to emerge from CAEM was that of "integral security," which considered economic development to be vital for Peru's security in the Cold War.

At the beginning, the idea of integral security was not incompatible with notions of Modernization Theory. Indeed, they very much resonated with the aims of American foreign policy in the first half of the 1960s.[96] However, ideas of integral security in CAEM began to change as the armed forces decided that foreign control of the economy was also a potential threat, as it stifled national economic development. Modernization Theory and its application during the Alliance for Progress had shown that while local capitalism was needed, foreign capital was also essential, so the Peruvian government had not tried to end American economic privileges in the region. Hence, the armed forces developed an affinity for Dependency Theory, as the generals believed that Peru's underdevelopment was the result of the development of the countries in the Northern Hemisphere. In this intellectual

FIG. 18. Building the Mantaro dam, holding back the waters. Corporación de Energía Eléctrica del Mantaro, *Memoria Anual 1968* (Lima: CORMAN, 1968). Biblioteca Nacional del Perú.

context, electricity—and hydroelectricity in particular—had now acquired a "strategic value," and rather than being seen as necessary infrastructure to promote economic growth, it had become a tool to fight underdevelopment. However, unlike previous governments, the GRFA would not seek to fight underdevelopment solely through infrastructure, but by pursuing an ambitious ISI program. Electricity became a key variable in modifying Peru's economy and unequal economic relations with the rest of the developed world, but not because it would create industry, but rather because it would accompany it.

The new regime expressed its disdain for previous liberal attempts at economic development only through infrastructure when it stated that the 1957 National Electrification Plan "lacked an integral understanding of the energy sector that would allow for the coordinated development of electricity with other subsectors."[97] As a consequence of this liberal vision, plants had been built without taking into account other economic activities, which had led to the "overabundance of electricity in some sectors, and a deficit in others." It concluded by stating that the main characteristic of the plan was merely an "inventory of existing and future projects."[98] The military's new development plan considered the relationship of energy with mining, agriculture, and industry and thus represented an integral development plan, a "fundamental output" that Peru had lacked in its "prerevolutionary context," in part because of inadequate legislation.

Such events were greeted with anxiety by the private sector, who as early as 1969 sensed that the wind was blowing in the direction of greater state intervention. A forum was organized, in which members of some of the country's most important electric companies took part, to promote the merits of the existing electric law. Although there was a consensus that greater public investment was required, electric businessmen nevertheless considered that "the state cannot assume all responsibilities without leaving other infrastructural developments unattended." Furthermore, the 1955 law, for all its liberal failings, had improved the distribution of electricity and had lowered costs. If the state were to intervene, it should do so only in a regulatory manner, such as overseeing the interconnection of a national system.[99]

Appeals by the private sector went unheeded by the generals, who in 1972 promoted a new electricity law stating that the supply of electricity constituted a strategic instrument necessary for Peru's "structural transformation."[100] The new legislation reserved for the state the right to generate, transmit, and distribute electricity through a new public electric company, ELECTROPERU.

The GRFA stated that the 1955 law was a ploy to keep electricity in the hands of the private sector, reducing it to a "utility and not an activity." Guided by its for-profit motive, the industry had developed unequally throughout the country, strengthening the country's centralization, as only Lima and a handful of other cities enjoyed an adequate supply of energy, while "our humble highland towns, our coastal farmhouses, and the small villages of our rainforests did not have access to the most elemental electric service." Had it not been for the actions of the GRFA, such a "situation would have continued, without goals, objectives, plans, achievements, studies . . . there would have been no electric conscience."[101] In a twist of fate, figures like Manuel Prado and Jorge Grieve, who had been the first to emphasize the need to develop an "electric conscience," now found their own words being used against them and their legacy.

With Mantaro still under construction, the legislation had no teeth. In the meantime, the military needed to maintain good relations with the private sector, particularly LL&P. LL&P's manager, Carlos Mariotti, also strived to stay on the government's good side. Mariotti had founded the IPAE-Escuela de Empresarios and its influential forum CADE (Conferencia Anual de Ejecutivos), which in the 1960s included figures ranging from the pioneer of Dependency Theory Raúl Prebisch to the proto-neoliberal Pedro Beltrán. Mariotti would use the forum to appeal to the military, hoping to promote linkages between them and Peru's industrialists. In private correspondence with the State Department, he stated that the creation of ELECTROPERU was a good idea, and that it would not mean that "private companies would be taken over by the government." American officials stated that LL&P's influence on the increasingly statist government was a tribute to "Mariotti's highly successful projection of Peruvian image (in substance as well as in appearance) of himself and his company."[102] It remained to be seen how long the company could hold their position.

While Mariotti's skillful maneuvering certainly aided LL&P during the early years of the revolutionary government, the company's existence hinged on the fact that Lima's power needs were ensured until 1970, and Mantaro would not come into operation until 1973. In the meantime, and not without considerable difficulty, LL&P inaugurated another plant in its electric pyramid when the Matucana Power Plant began operations in 1971. Producing 120,000 KW and once again funded by the IBRD, Matucana was named after Pablo Boner, the company's longtime electric architect, who died in the year following its inauguration. The company likewise submitted to the IBRD a final proposal

FIG. 19. Electric nationalism. *Electrotécnica*, no. 53 (October 1973). Asociación Electrotécnica Peruana.

to build the Sheque Plant—also designed by Boner—to be constructed above Huinco, but Matucana proved to be the last plant built by the LL&P.

After the inauguration of Matucana, LL&P began to realize that, despite Mariotti's influence, the company would not be spared from the government's nationalist rhetoric. In late 1972 the government decreed that all foreign assets of electric companies were to be acquired by the state. This presented serious problems for LL&P, for although Swiss capital had reached its zenith in the 1950s and 1960s, by the 1970s 25 percent of its shares were still controlled from Zurich. In time, almost half of the company's shares were acquired by the government, but the latter still did not feel confident enough to nationalize the company, assuring LL&P that they would continue to enjoy their "traditional rights and obligations."[103]

Meanwhile, in Tayacaja, the machine house was being finished, and the first turbines were being installed. The machine house, embedded in the mountains of Tayacaja, would not have seemed out of place with the *andenes* that Ráez y Gómez had witnessed at the beginning of the century. Finally, on October 7—celebrating the five-year anniversary of the revolution—with already half its generating capacity installed, the government inaugurated the plant. Despite the revolutionary fervor, the plant's inauguration was eerily reminiscent of that of Callahuanca, Peru's first large hydroelectric plant, almost forty years earlier. Much like in 1938, the site was dominated by Italians, with an estimated "one hundred high functionaries of the Italian government, representatives of industry and commerce."[104] A special flight had been scheduled from Rome to Lima for their arrival, and among the dignitaries was Dionigi Coppo, an influential figure in Italy's Christian Democratic government. Likewise, the military ordinariate of the armed forces, Monsignor Alcides Mendoza, blessed the site, much as the apostolic nuncio had blessed Callahuanca. Even the Swiss, who had once again provided technical oversight during the construction, were in attendance. Finally, Antúnez de Mayolo's son was invited by the authorities. One may wonder what Antúnez de Mayolo, a vocal supporter of land reform, would have made of this peculiar regime.[105]

While continuity and tradition coexisted amid the revolution, the language had clearly changed. General Fernández Maldonado stated that if Prado had wished to maintain old divisions, the GRFA wished to abolish them. While Belaúnde hoped to "reconquer" Peru, he did not embark on the real conquest that the country needed, ending imperialism and dependency. Indeed, the general stated that the Mantaro project, as used by the previous

FIG. 20. Mantaro Hydroelectric Plant embedded in the mountain. Ministerio de Energía y Minas, *Mantaro* (Lima: Ministerio de Energía y Minas, 1973). Biblioteca Nacional del Perú.

"macrocephalic Lima based governments," had not allowed the country to exploit its true potential. Hence, the revolutionary government found the plant in 1968 in a "complete state of abandonment, agony, in a process of extinction; the spoils of its construction going to large international firms." It was at this critical moment that the GRFA intervened, giving CORMAN the necessary human resources, financial backing, and, most important of all, a "revolutionary mystic."[106] The plant's inauguration was also used to attack the enemies of the revolution. Fernández Maldonado celebrated the "loyal" Italian foreign investment, as opposed to the "freebooting, deceiving, and false actions of other foreign companies, who have written and continue to write the pages of oppression against the interests of the Peruvian people," directly alluding to the IPC, "forever expelled from the country."[107] If oil had been historically perceived as a fertile ground for imperialism, hydroelectricity represented a new age of nationalism.

General Velasco, who remained in Lima—he did not want to overshadow Gen. Fernández Maldonado and upset the delicate balance of the "institutional" military government—sent a celebratory message to the workers of Mantaro. No longer was he congratulated for being a mythical "Andean" figure who thrived in harsh conditions, he had now become an iconic figure in the revolution. "The workers . . . have given their quota of sacrifice . . . The fallen, are already a part of the history of the development of our fatherland. The construction of this hydroelectric plant shows the unbreakable faith of Peru in the technical, operative and revolutionary capacity of its sons."[108]

One year later, the GRFA published a pamphlet that developed the revolutionary government's conception of hydroelectricity and geography. Mantaro was especially important for uniting a territory that was more like a "country-continent" than a country because of its many races, many languages, and "pluralist geography." Indeed, the geographic element, so prevalent in previous governments, could not be ignored by the revolution. Fernández Maldonado stated that the Incas had been able to thrive in the area because of their socialist inclinations, but modern Peruvians continued to face a "singular geographic reality that posed a great challenge to . . . any development effort with clear goals." Further accentuating this reality between possibility and challenge, the government realized that Peru "has all the riches of the world that in other cases are found in countries with much larger territorial extension."[109] The Mantaro Plant, which allowed for harmony between this challenge and possibility, formed the cornerstone of Peru's revolution, as it permitted, in a phrase reminiscent of Lenin, "Revolution with Electrification."

With the first phase of the project completed, the government proceeded to nationalize the entire electricity sector. Since their arrival in 1968 the military had considered electricity to be essential to develop Peru's industrial potential and, equally importantly, to light the houses of newly arrived immigrants to Lima, who, by 1970, accounted for 25 percent of the city's population.[110]

As David Collier has pointed out, the military government thought that residents of the barriadas had accomplished a great deal with little support from the state.[111] But no matter how self-sufficient Lima's new inhabitants were, they did not have the capacity to generate electricity by themselves. During Manuel Prado's government, despite his desire to reduce migration to the city, thousands of Lima's residents received electricity thanks to the efforts of Junta Nacional de Vivienda and, most of all, to LL&P.[112] Under Mariotti's leadership, since the 1960s, the company had developed a progressive policy of providing electricity to the barriadas and had financed many of the connections. This greatly enhanced the reputations of the company and its manager.[113]

The GRFA was to take a far more aggressive approach than its predecessors. Not sharing oligarchic anxieties of an old aristocratic Lima fading away, the government did away with the pejorative term "barriadas" and started using the term "*pueblos jóvenes*" (young towns), which had been coined in 1967 by Bishop Luis Bambarén.[114] The new name also symbolized a new attitude toward urban growth. Unlike Prado, who saw the barriadas as an "emergency" condition that was to disappear with the development of the interior, the term "pueblo joven" implied that these new urban spaces were to mature and become part of the city. With over a million inhabitants by now, the pueblos jóvenes were seen as ripe for cultivating nationalist and "revolutionary" support for the government. Through its Office for the Development of Pueblos Jóvenes (ONDEPJOV) the government sought to connect the shantytowns to the electric grid, not only to improve living conditions but also because it believed that electricity was essential to reach these populations politically via television and radio.[115]

Soon, the inhabitants of the shantytowns directly appealed to the revolutionary government for electricity and found in infrastructure a way of expressing their citizenship. Mario Julca, an inhabitant of Villa María del Triunfo, wrote to the Ministry of Energy and Mines asking to be supplied with electricity as it was not only a "primordial necessity of the home" but, more importantly, a matter of "justice."[116] Inhabitants of a housing cooperative—aptly named Tayacaja—begged for light to reduce crime in their neighborhood, a right they were entitled to "as good revolutionaries and cooperativists working for the welfare of the nation."[117] Residents of Villa El Salvador asked to be provided with light for the 27th and 28th of July, so that they could properly celebrate the national holidays.[118] Unfortunately for the military government, all these requests continued to be forwarded to

LL&P, which was still in charge of the grid. However, the military no longer wished for an intermediate entity between the government and its people, and soon ONDEPJOV would take direct action. No matter how progressive its policies were, LL&P's days were numbered.

Thus, in 1973, Mantaro became interconnected to the LL&P system. The 220-volt transmission lines rose from Tayacaja, climbed over the central Andes, and descended upon Lima. The grid was connected to Santa Rosa—LL&P's first plant in the city—where the company's control board was located.[119] Soon, the electricity of Mantaro flooded the region, engulfing everything in its path and finally creating Peru's first truly interconnected infrastructural grid. The GRFA made its final assault. The Cerro de Pasco Corporation was nationalized in 1974 and became CENTROMIN. Likewise, the GRFA decided to not renew LL&P's concession rights over the city of Lima, putting an end to a monopoly that had lasted for over half a century. The company was nationalized and renamed ELECTROLIMA, and Carlos Mariotti departed to Switzerland. Cooperatives were also not spared, as the Cooperativa Comunal del Centro had been, but this happened later—it would be absorbed by 1976. The revolutionary fervor put an end to García Calderón's beloved trust and Mariátegui's cooperative alike.

TABLE 2. INSTALLED CAPACITY, 1968–1980 (MW)

	1968	1970	1972	1974	1976	1978	1980
Hydraulic	915.1	922.6	1056.8	1338	1397.3	1408.8	1864.7
Total	1606.5	1677.1	1930	2265.7	2473.3	2570.3	3184.3

Source: Ministerio de Energía y Minas, *Minería y energía en el Perú: Política y situación actual* (Lima: Ministerio de Energía y Minas, 1977); Pedro Pablo Kuczynski Godard, *Memoria de Energía y Minas 1980–1982*, Tomo 1 (Lima: Ministerio de Energía y Minas, 1982) and *Siete años de revolución en el sector energía y minas* (Lima: Ministerio de Energía y Minas, 1976).

As Lima's poorest began to receive electricity, it became clear that Prado's "great exodus" was forever doomed. But Velasco and his generals would also be defeated. For them, Mantaro was meant to break Lima's dominance over the rest of the country. Yet by the time Mantaro started to supply power, Lima had received so many Andean migrants that no other city could compete with it either demographically or economically. And the only solace that the desolate shantytowns could find in the urban chaos was access to electricity.

In the end Mantaro reinforced Lima's position, much to the dismay of revolutionaries, but it was not the Lima of the past, much to the dismay of the oligarchs. Indeed, the "Lima of one million" of which oligarchs anxiously spoke at the beginning of the century would eventually become a city of over ten million inhabitants powered by the energy of the Andes.

CONCLUSION

As the "Peruvian experiment" of the GRFA continued, divisions between radicals and moderates within the institution began to intensify. Morales Bermúdez, no longer the young colonel working in CORMAN but now second in command as minister of economics and finance, carried out an internal coup against an ailing Velasco in 1975. This marked a new phase in the revolutionary government. Land reform ground to a halt and diplomatic relations with the United States began to improve, although relations with the Soviets continued. ISI policies were abandoned, as many of the nascent industries proved to be unsuccessful and highly expensive, burdening the Peruvian state with a massive foreign debt for years to come. As had happened with previous governments, the only semblance of continuity between the two phases of the revolutionary government seemed to be the construction of electric infrastructure. After all, Lima kept growing, and Mantaro still had to undergo a second phase of expansion. The total national output by the end of the revolutionary regime was over 3,000 MW—with almost 2,000 being hydraulic—but the century-old developmentalist aspirations that electricity had represented remained unfulfilled.[120]

Designed by Peru's foremost engineer, Santiago Antúnez de Mayolo, Peru's largest hydroelectric plant represented a clear case of "scientific excellence from the periphery," as the "wiseman" creatively applied Western scientific practices on a challenging Andean topography. Informed and inspired by ancient cultures, Antúnez de Mayolo's endeavors in many areas represented a rediscovery of how Peruvians could thrive in the Andes. The new technology fit comfortably in the mountain chain, as Antúnez de Mayolo saw his actions as a continuation of Incan infrastructural grandeur, which had been lost since the Spanish Conquest. The same could be said of those involved in the physical construction of the project. Once again the revolutionary government characterized these workers as the only actors who could have carried out such an ambitious feat although, despite its radicalism, this regime celebrated them while keeping them anonymous, just as previous governments had done.

The comings and goings of the Mantaro scheme show how infrastructural projects were characterized by physical continuity and, equally so, by political ruptures. Manuel Prado could not have imagined the new ideological meaning that the revolutionary government would give its beloved political project. The military, in turn, could not comprehend Prado's oligarchic yearnings. For elites in Lima, Mantaro symbolized national development, for regional elites, the end of centralized control from the capital. The Mantaro Plant could strengthen the international relations of Peru with its "natural" Western European allies, or it could be used to approach the Soviet Bloc, which had been largely avoided by previous administrations. Electricity, a seemingly technical and objective enterprise, was politically fluid and versatile, representing different social and economic goals according to the ideological leaning of those in power. But while the project divided Peru's political class, it simultaneously connected the country, as its culmination symbolized the establishment of Peru's first-ever interconnected grid.

In 1979 the last turbines were installed, and the plant was finally producing the 800 MW envisioned by Antúnez de Mayolo. The final step in his vision was the construction of the second plant below the one named in his honor. In 1984 this became a reality, and another 210 MW of power were added. Mantaro was finally producing over 1 GW of energy. The name of the second plant—Restitución—symbolized returning to the Mantaro River the waters that the state had taken from it. And, as the military went back to the barracks in 1980, it also marked the return of democracy. Some time afterward, electricity, which had been appropriated by the state, was restituted to the very civil society from which it had emerged. Like many of the revolutionary government's other reforms, the public nature of electricity survived throughout the 1980s but was not able to withstand the neoliberal onslaught of the 1990s: as the grid was privatized, the country's electric landscape once again became fragmented. As the state retreated into the shadows, Mantaro proved to be the end of Peru's infatuation with hydroelectric infrastructures.

Epilogue

FROM LIGHT TO DARKNESS

As the Revolutionary Government of the Armed Forces came to an end, so did the era of Peru's long-sought dreams of industrialization. Hoping to claw its way out of a severe economic crisis, the government halted and even reversed reforms. The economic decay was such that in the final years of the revolutionary government energy consumption in Lima stalled for the first time in the century. Despite political and economic instability, the planning frenzy that had begun in the 1950s could not be stopped, having been thoroughly incorporated into state dynamics. With the assistance of the Federal Republic of Germany and the World Bank, the regime crafted a new electric plan, which was meant to guide the development of the sector for the 1980s. As with previous plans, priority was given to hydroelectric projects.[1] Engineer Azi Wolfenson, new head of ELECTROPERU, inherited a public company whose debts exceeded its capital, but by the early 1980s it had improved its financial position. With Lima's demand for energy again on the rise by the beginning of the decade, ELECTROPERU began to dust off decade-old projects. These included Pablo Boner's plans to build a hydroelectric plant, above the Sheque Reservoir, that fed the Huinco Power Plant, pushing once more the vertical limits achieved by Lima Light and Power (LL&P) in the 1960s. According to Wolfenson, the project would have increased Lima's power supply and satisfied its water needs. Even Santiago Antúnez de Mayolo's plans to redirect the waters of the Mantaro toward Lima were debated once again.[2]

While such plans would have been eagerly supported by previous administrations, it soon became clear that the enthusiasm for hydroelectric projects was not shared by Fernando Belaúnde Terry, who, after his long exile, was again elected president in 1980. The once enthusiastic and optimistic architect no longer sought to reform Peru's social and economic structures. Acción Popular (AP) had ceased to espouse any of the reformist tendencies that it had championed in the 1960s. Its newest incarnation was eager yet unable to fully embrace the new neoliberal currents that had arrived in Latin America.[3]

AP's shift to the right was also reflected in the types of infrastructure promoted by the government. Belaúnde continued to superficially speak of an infrastructural nationalism, but aside from his road-building initiatives, ambitious projects were put aside. The arrival of a young Pedro Pablo Kuczynski Godard—future president of Peru—as the new minister of energy and mines cemented the government's new policy. In his first annual report, the new minister used a familiar phrase to describe the state of the electric industry. "In electricity, the saying attributed to Raimondi, that Peru is like a 'beggar sitting on a pile of gold,' is a faithful description of our reality." But rather than using Raimondi's words to push the state to exploit the country's hydraulic riches, Kuczynski now employed them to criticize excessive government intervention. The "beggar" was Peru's own electric infrastructure, which, despite making great strides throughout the century, by 1980 provided electricity to only roughly 40 percent of the population, the majority of whom inhabited urban centers. Kuczynski placed the blame on the "proverbial inefficiency of the public electric company and an asphyxiating centralization." The "pile of gold" continued to be Peru's hydraulic potential, but hydroelectric projects were deemed too difficult and expensive to develop. The "pile of gold," thus, "is a relative satisfaction, and taking advantage of it will not be easy."[4] Kuczynski's words symbolized a clear break with previous governments, for whom hydroelectrification was considered a priority to solve Peru's underdevelopment.

Despite this new neoliberal outlook, the new government was unable to reverse many of the "revolutionary" policies of the past decade. Regarding the oil sector, a new law—infamously known as the "Kuczynski Law"—once again offered generous tax breaks to foreign firms to enter the field and showed Kuczynski's predilection for oil as opposed to hydroelectric technologies. The law proved so unpopular that it was repealed toward the end of the government. As for hydroelectricity, the government could not pass such ambitious legislation. When a new electric law was passed in 1982, it emphasized that electricity was a public good that was to be provided by the state. The law, however, sought to increase the participation of the private sector, allowing small-scale producers—mainly mining companies—to expand their facilities and sell excess generation to ELECTROPERU. Clearly unhappy with the limitations of the law, Kuczynski warned that "only the future would tell if this moderate position is the right one," hinting that the state should sell some of its investments, although he admitted that the wholesale of plants was at the moment "politically difficult."[5]

Regardless, hydroelectricity—and, by extension, the state—seemed to no longer hold the promise that it did in the past. So too did the influence of its advocates. The great engineers of the twentieth century had passed away, and the politicians who had supported them had also died or retired. Technical/lobby groups, such as the Peruvian Electrotechnical Association (AEP), entered a process of decline. Its publications, which since the 1950s had included articles celebrating the link between electricity and development, became dry technical pieces on the workings of electricity, illegible to the lay reader. In the past the AEP had been able to summon statesmen and industrial leaders, but as time passed the corridors and conference rooms of its once modern locale in downtown Lima became empty.

1980s Peru not only suffered a deep economic crisis that was shared by most Latin American countries, in what came to be known as the continent's "lost decade," it also faced the direst crisis of its political history. As the opening lines of the book detailed, the Shining Path—or Sendero Luminoso in Spanish—began its activities in 1980, and it lost no time as it moved into the central Andes to attack the region's electric infrastructure. Much like the geographers of the early twentieth century, *senderistas* were well aware of the importance of the central region to the capital. The Mantaro Valley, with its high population density, its considerable agricultural production, and its strategic position as a corridor between the coast and the Amazon, became a key battleground in Peru's internal conflict.[6]

The inauguration of the Mantaro Hydroelectric Plant in Huancavelica only added to the strategic importance of the region, as its electric grid crossed the Mantaro Valley to reach Lima's households and industries. On December 13 the Shining Path blew up a high voltage tower near the town of Concepción, one of the first armed actions in the valley.[7] After their opening salvos, the situation escalated rapidly, and after they realized the psychological impact of blackouts, attacking the electric grid became the Shining Path's preferred strategy. While only five towers were felled in 1980, by 1985 the group had become particularly adept at creating darkness, bringing down over one hundred towers that year alone. Two years later, on New Year's Eve of 1987, the largest attack against the Mantaro interconnected electrical system took place, with over thirty high voltage towers blown up.[8] While Sendero's main purpose was to bring darkness to the capital, the inhabitants of the Mantaro Valley, who had enthusiastically dreamed of electrification when developing their electric cooperative in the 1960s, also discovered the terror of losing the electric power they had recently grown accustomed to. A particularly heavy blow was dealt when the valley's milk processing plant—considered the greatest industrial achievement of their electrification efforts of the 1960s and

1970s—was blown up. Soon afterward, the people of the Mantaro Valley would take up arms to defend themselves from the terror inflicted by senderistas.[9]

The Shining Path also tried to capture the Mantaro Plant more than once, the first attempt taking place in November 1981 when Campo Armiño was attacked, although Sendero was swiftly repelled by Peru's Republican Guard—Peru's national state gendarmerie.[10] The importance of the Mantaro Plant meant that the small city of Pampas, in the province of Tayacaja, became one of the most important operation centers for the counterinsurgent activities of the armed forces.[11] The attacks against the grid and the Mantaro Plant also led to an unexpected development in the Peruvian state, which was the creation of the Energy Security Force (FUSE) within Peru's Republican Guard.[12] FUSE specialized in seizing Sendero's dynamite, guarding electric towers, and defending the Mantaro Plant from armed attacks. Once the revolutionary attempts of the early twentieth century had been left behind, infrastructural politics had been disputed through policy and planning, but now they once more became a physical battleground as the state and insurgents fought one another for control of the electric grid.

Soon, Lima, oblivious to events in the southern and central Andes, became aware of how much it depended on the Andes for its power. News reports would highlight the vulnerability of the grid by simply stating that the connection with the Mantaro Plant had been shattered. In July of 1983, for instance, Lima's main daily reported that "six electric towers were dynamited last night simultaneously by terrorists which led to a general blackout in Lima," with ELECTROLIMA describing the attacks "as the largest against its electric services." The greatest damage was inflicted upon tower no. 131 of the Moyopampa–Lima transmission line, which had been built by LL&P in the 1950s and subsequently connected to the Mantaro system. The promise of infrastructure seemed to come crashing down with the transmission tower, as "three out of the four legs gave way after the dynamite charge" and the structure, which held lines of six thousand volts, ended up "practically on the ground."[13]

A blackout meant not only a night of darkness for citizens but also a grinding halt to all the industrial activities that electricity was supposed to power. It also meant being deprived of other essential services. After that same attack, "many factories stopped their production, a good number of radio stations could not transmit their signals, three national newspapers could not go to press, gas stations could not provide fuel, and many sectors of the city had to go all day and night without the essential services of electric light and drinking water."[14] Unable to provide electricity, ELECTROPERU had to draw power from the northern part of the grid, which was connected to the Cañón del Pato Hydroelectric Plant. Designed in the 1940s for the Santa Corporation and meant to industrialize northern Peru, the plant now had to provide light to Lima as well.[15]

No place in the capital was spared. Lima's shantytowns, which had gradually been electrified since the 1950s, became prime targets for the Shining Path. In 1983, a year in which the attacks in Lima began to intensify, San Juan de Miraflores—which originally emerged in 1954 as a barriada named "Ciudad de Dios," before being categorized as a pueblo joven by the revolutionary government—suffered "a new synchronized attack" when ten transmission towers were blown up in a single night, explosions taking place within ten minutes of one another. The subsequent blackout allowed terrorists to attack other parts of the city, including the police headquarters in Chorrillos.[16] Many were injured, and, for those who had been alive in the 1930s, these events were reminiscent of tactics used by APRA, although Shining Path proved to be more successful in their actions. That same year Sendero began attacking embassies and electrical installations in the city's wealthier neighborhoods. However, while such news dominated the front pages, little was shared about the atrocities that the Shining Path inflicted on Andean peasants, atrocities that would be replicated by the armed forces as they unsuccessfully tried to eliminate the Maoist terrorists.

The development that electricity had seemed to promise began to be rolled back. First, there was the cost of repairing the towers, which averaged thirty thousand U.S. dollars per tower and, in some cases, took a month to complete. The cost of repairing all towers felled in the 1980s was over 24 million U.S. dollars, with an additional 13 million U.S. dollars spent in fuel for the operation of emergency equipment, as well as 40 million U.S. dollars in lost energy sales.[17] The overall damage exceeded 77 million U.S. dollars, close to half the amount that had been invested in the construction of the first phase of the Mantaro Hydroelectric Plant, and 50 million U.S. dollars more than had been spent building Huinco.

For those Peruvians who lived through the armed conflict, terrorism was synonymous with darkness. But electricity played a key role in other ways. As Sendero targeted community leaders in order to create political vacuums that they sought to fill, they murdered the head of the electrification committee in Sacsamarca, Huancavelica.[18] At times, they also tried to influence the population through controlling such organizations, as in the case of Huaycán, a pueblo joven organized and promoted by Izquierda Unida (United Left), and whose electrification committee was infiltrated by Sendero to weaken their political enemies.[19] Engineers were also involved in Sendero's ruthless campaign. Through dynamics of "popular aid," the group sought out potential senderistas that were "more useful in their present occupations than they would [be] as full-time militants."[20] Among these professions were electrical engineers, in charge of tipping off Sendero on the most vulnerable sections of the electric grid. Even the United Press reported that military sources had been complaining that the "rebels had infiltrated the Armed Forces, the police, and the strategically important state electric company."[21] In short, Sendero Luminoso's fight against the Peruvian state would have been dramatically different had electric infrastructure not been in place.

The actions of the Shining Path during these years highlighted the problems presented by an interconnected grid in times of armed conflict. Early-twentieth-century revolutionaries and apristas of the 1930s had to capture Lima's plants directly if they wished to sow chaos through the production of darkness. However, half a century later, the Shining Path could blow up a lone tower in the middle of the highlands, and the consequences would be quickly felt in the capital. The creation of an interconnected grid not only highlighted Lima's vulnerability but also symbolized a profound paradox. Most limeños had paid scant attention to the growing power and influence of the Shining Path and saw its actions in the Andes as distant events that had no impact on their daily lives. Such blissful ignorance was shattered when the darkness came to the capital. Throughout the twentieth century, Lima became increasingly connected to the Andes as the grid was being built, becoming dependent on the energy created by the Andean hydraulic machine. But the same modernization had not resolved—and had perhaps even contributed to—the disconnect between Lima and the Andes, which regional intellectuals had been lamenting since the beginning of the century.

Each blackout added to the public perception that the country was on the verge of becoming a failed state, even though the electric grid was meant to symbolize its infrastructural power. The blackouts meant that old diesel engines were revived, and, for a short time, the capital once again consumed a considerable amount of thermal energy. It is ironic that the Maoist terrorists, in their fanatical search for Mariátegui's "shining path towards the revolution," had to plunge Lima and the central Andean region into darkness.[22] Electricity, which had just recently unified once disparate territories, had become the Achilles' heel of the Peruvian state.

Perhaps more critically, Sendero's actions once more changed the coastal perception of what the Andes represented. For a good part of the twentieth century, the Andes seemed to no longer be an "obstacle" for coastal elites, but an asset, thanks to its hydroelectric potential. One of the many tragedies of the arrival of the Shining Path was that the Andes once more came to be viewed with pessimism—a region that, instead of modernity, produced violence—seemingly confirming the elites' inherently negative view of the region. Throughout the twentieth century the prophets of light had prevailed, and now the Maoist prophets of darkness seemed to take center stage. Such developments, however, did not push Peru's elites to reconsider their relationship with the Andes, but simply seemed to cement old racial divides that had plagued Peruvian history.

The Shining Path's fight against the Peruvian state would end in failure, but other events dramatically transformed the nature of the electric sector. In 1990 an unknown candidate by the name of Alberto Fujimori won the presidency, signaling the arrival of neoliberal populism in Peru. The pendulum, which throughout the twentieth century had swung in the direction of greater state intervention, now wildly swung back against the state. In 1992 a new law was passed that liberalized the electricity sector, privatizing every plant near Lima. The Mantaro Power Plant, crown jewel of Peru's electric system, proved too big to be given away to private interests, so it remained in the hands of the state. However, the new law did modify how the power it generated was transmitted and consumed. Previously, all electric activities—generation, transmission, and distribution—had been under the control of a single entity. Such had been the case with LL&P, CORMAN, and eventually ELECTROPERU. Now each activity was separated from the others, and concessions granted to private interests.[23] Although still connected, in many ways the grid had been split apart.

Some continuity can be found in times of change, as the final step toward the creation of an interconnected national system was made almost forty years after Manuel Prado approved Peru's first decree seeking to interconnect Peru's disparate electrical grids. Throughout the 1990s two main systems were established, ETECEN (Sistema Interconectado Centro Norte) and ETESUR (Sistema Interconectado Sur). In 1998 preparations were made in Campo Armiño in Huancavelica and the Socayaba Power Plant in Arequipa to connect the two systems. In July 2000 the grid was finally unified. Most Peruvians were not aware of this momentous event, and most experienced only a small glitch in their television screens.[24] That same year, Alberto Fujimori would be reelected for a controversial third term. While the populist president would stay in power less than a year, as corruption scandals pushed him out of office, his presidency deeply polarized the country. Electricity had become the single infrastructure to truly unite Peru's territory, but, politically, Peruvians remained as divided as before.

Although Peruvian developmentalism has come to an end, the story of Peru's hydroelectric sector is far from over. It now faces three critical challenges. The first challenge is one that the architects of Peru's electric system could not have foreseen in their lifetimes, that of climate change. Certainly, Peruvian and foreign engineers had always dealt with the wrath of nature. All the electric plants in the central Andes had been damaged by landslides, or *huaicos*, throughout the century—indeed, they had been designed to withstand such natural disasters—but by the beginning of the 1980s nature's wrath intensified. The consequences of *El Niño*, a natural phenomenon that takes place off Peru's coast, became more dire as the century ended. Landslides or rising rivers halted the activities of Callahuanca, Mantaro, and various sectors of the grid.[25] Much like the fallen towers, however, they were rebuilt. Callahuanca, which was last impacted by a destructive landslide in 2017, resumed operations less than two years later. It continues to provide light to around 500,000 families after almost one hundred years of continuous operation, a fine example of the sophistication and quality of the engineering work of its creators.[26]

While Peru's hydroelectric infrastructure may withstand the impact of natural disasters, it is uncertain if it will survive the long-term effects of global warming. On the one hand, as the Committee on Climate Change (COP) has emphasized, hydroelectric energy is a renewable resource that can be used to fight climate change. On the other hand, it is probable that the availability of water for the generation of energy will decrease, as precipitation patterns become more unpredictable and the glaciers of the Andes continue to melt.[27] Engineers in the past never complained about a lack of water—there were certainly dry spells, but these were seen as temporary—but of the difficulty of gaining access to it. Reginald Enock described the Andes as a "hydraulic machine" over a century ago, but it remains to be seen if the hydraulic generosity of the Andes will continue to manifest itself in the coming years. If the effects of climate change are not mitigated, the hydroelectric plants built high in the Andes will become relics, no different from the great Inca ruins that led engineers and politicians to celebrate the harnessing of the Andes' vertical nature.

The second challenge relates to the international domain, as the geopolitics of energy are shifting. Throughout the twentieth century, European interests—mostly Swiss and Italian—dominated the development of the sector, reflecting existing patterns of technological and capital flows. By 2024 the China Three Gorges Corporation and the China Southern Power Grid International had acquired all of Lima's electric infrastructure, including the iconic plants built by LL&P throughout the century. While the United States has expressed its concerns, unlike in the years of the Second World War, Americans seem unable to counter external influence in the region.[28] While it is still unclear what the political consequences of this investment will be for the Peruvian state, it once more demonstrates that global battles between great powers continue to be fought in the infrastructural arena, even if the actors have changed.

The third and final challenge continues to be Peru's political instability. From 2016 onward Peru entered a period of political turmoil characterized by renewed demands from Andean citizens for greater state presence, as the neoliberal model implanted in the 1990s showed its limitations. In 2019, residents of Huancavelica, amid protests for a greater economic benefit for their potato harvest, burned Campo Armiño to the ground. Previously they had enjoyed cordial relations with plant workers, used the plant's health services, and received electricity at very low rates. But the attack took place precisely because the plant remained the only symbol of state presence in the region, so there was no other physical structure to protest against. The Mantaro plant was targeted again in 2022, the deadliest year since the armed conflict, as President Pedro Castillo, an indigenous rural schoolteacher whose support came mainly from Andean provinces, was ousted from office after a failed coup d'etat.[29] As Peruvian politics continue to deteriorate, armed criminal gangs related to illegal mining have once again taken to blowing up electric towers in the Andes to attack large mining firms. The lack of ideological substance of such actions matters little to the distant citizens of Lima, as they continue to distrust the Andes and its inhabitants. The Andes are connected to Lima by the electric grid but remain separated in every other possible way.

The events of the 1980s and 1990s brought into dramatic relief that decades of modernization efforts, through large-scale national electrification projects, had failed to bring a much-coveted national social integration or political stability. A historical approach analyzing the nature of infrastructural politics shows that the development of what political scientists have termed "infrastructural power"—at least in its physical manifestation—is anything but simple. Political scientists have demonstrated that the expansion of bureaucratic entities and the construction of different forms of infrastructure expand the power of the state but, surprisingly, seldom consider the actual *politics* behind these complex processes. While the construction of infrastructure no doubt expands state power, it may likewise pose considerable challenges as projects become politicized.

In the Peruvian case, these divisions clearly manifested themselves with the very arrival of hydroelectricity. Since the beginning of the twentieth century, intellectuals believed that hydraulic sources of power could point the way toward greater political decentralization and provincial autonomy. Others—such as Manuel Prado, later in the century—thought that hydroelectricity could maintain Lima's dominance over the nation as existing political and social hierarchies were increasingly questioned. Hydroelectricity could also reinforce democratic values or reinforce authoritarian aspirations, depending on the ideological leaning of each administration as well as the political climate of the time. It could maintain tradition under the guise of modernity, or it could equally represent a revolutionary endeavor to destroy the traditional order. These different meanings emerged not only in different infrastructural projects, but sometimes within the *same* project.

In a nutshell, the violent political comings and goings of the Peruvian Republic were reflected in its very infrastructure. Yet, in a context in which politics represented constant ruptures, as civilian and military governments alternated power between them, infrastructure more often than not was the only continuity between these disparate regimes. But the continuity was only physical—construction continued, electricity was consistently generated—the political meaning changing with the ebb and flow of Peruvian politics. Hydroelectricity, thus, was much more than the generation of electric power—it shaped notions of who could enjoy political power and how it should be deployed.

Infrastructural politics were also inevitably tied to Peru's geographical configuration, as any form of infrastructure is. But unlike roads, an infrastructure negatively affected by the country's topography, hydroelectricity ushered an element of ambiguity into Peruvian geography, depicting its generous nature. In many ways, the vision that most Peruvians have today of their geography is far more negative than that which emerged at the beginning of the twentieth century. Just like over a hundred years ago, many of Peru's ills continue to be blamed on its geographic configuration, but, unlike a century ago, there is little hope that this configuration may offer some developmental possibilities. For early intellectuals, the nature of Peruvian geography was never reduced to a simple debate on its positive or negative attributes. It was always regarded as complex, even if the national territory had been divided along three well defined "natural" regions. This complexity was centered around notions of verticality, which, for the first time since the days of the Incas, could be exploited thanks to hydraulic technologies. Today, the electric craze has largely disappeared from the public's imagination—infrastructures in general no longer inspire Peruvians as they did throughout most of the twentieth century—and its disappearance took with it any reimagination of Peruvian geography or the generosity of its vertical nature.

When speaking of geography, Peruvians were never simply referring to the physical territory. They were also speaking about Peru's different peoples. Just as the Andes were deemed to be both a problem and a possibility, Peru's Indigenous people were viewed in the same way. In the pre-Columbian past, Peru's Indigenous populations had thrived in the highlands, while in republican times, intellectuals portrayed them as being "passive" and pointed out the difficulties of incorporating them into the nation-state. But at the same time, both foreign and national observers noted that there was potential—the degree of which varied according to the writer—to be found in Peru's Indigenous peoples. García Calderón thought that there was some role to be played by "Indian" populations in a technical capacity. Others, like Mariátegui, considered them to be the soul of the nation, an idea echoed by both Belaúnde and Velasco—although the latter would speak of *campesinos* rather than "Indians."

How this potential was to be tapped was open to debate. This book reveals that the celebration of the man of the sierra or the altitudes found its best expression in the construction of hydroelectric projects, as it represented the right kind of infrastructure through which Peru's Indigenous peoples could once more reconnect with their almost mythical engineering traditions. In part, this was the result of Peruvians discovering that it was difficult to thrive in the Andes, even when harnessing its verticality. Much ink has been spilled regarding Inca ability to flourish in the mountain range almost effortlessly. Perhaps the construction of great hydroelectric plants showed that modern Peruvians—much like the Incas—had to work hard to exploit their national territory. Through such challenges, hydroelectricity allowed the Andean population and its territory to be connected once again in infrastructural grandeur. Thus, hydroelectricity was pivotal in reimagining the Andes, but these new imaginaries were grounded in the opportunities offered by both modern technologies and ancient practices. Such reimagination, however, could cut both ways, as it remains unclear whether Peru's Indigenous populations were merely "useful" in this endeavor, as García Calderón had prophesized, or essential, as a thankful Pablo Boner would celebrate. In the end, by geographically limiting the role of Indigenous populations to the Andes, and to one specific infrastructure, hydroelectrification not only celebrated past glories but cast a light on modern prejudices as well.

This book also demonstrates that despite Peru's weak bureaucratic state, one of the few Peruvian institutions in which a permanent body of bureaucrats existed was the electric sector within the Ministry of Fomento and Public Works, at least when it came to its Directorate of Industry and Electricity. Therefore, the development of physical infrastructure also had an impact on other variables of infrastructural power, such as the development of specialized state institutions. Indeed, Fomento officials came closest to the concept of a Weberian bureaucracy—rational, continuous, and with long-term plans. Despite their apparently apolitical nature, these bureaucrats were inevitably tied to the dynamics of infrastructural politics as they navigated the constant political changes that rocked twentieth-century Peru. As ministers came and went, these bureaucrats stayed in place, serving the Peruvian state for almost three decades, interacting with local and foreign scientists, national and foreign capitalists, international aid regimes, and local populations. They also established strategic alliances with scientists through technical bodies, successfully crafting electric plans and lobbying different governments to show a greater interest in the development of hydroelectric projects.

These efforts seemed to bear fruit. At first, hydroelectric development was completely overlooked by the state as private interests monopolized it, but eventually the state would push to control the whole sector. Of course, these bureaucrats were dependent on the whims of Peru's presidents, some of whom paid little attention to hydroelectric development and some of whom had an outright obsession with it. As the century moved forward, the alliance between the state and science, which had begun in the early days of the Aristocratic Republic, found its ultimate expression in the construction of Peru's colossal hydroelectric plants.

A final word should be said about the transnational linkages that characterize the nature of infrastructural power. Political scientists, no doubt interested in the state, tend to analyze its internal dynamics, leaving aside that the development of complex infrastructures tends to require considerable foreign aid and technology, especially in countries that, in Gerschenkronian terms, are playing "catch up." Thus, governments wishing to embark on ambitious hydroelectric projects had to navigate not only complex domestic political dynamics, but also the international context that determines the flow of capital and technical expertise.

These transnational linkages were also politicized. In the first half of the twentieth century, private interests tended to dominate the electrical sector, while the Peruvian state played a secondary role through the establishment of educational institutions. However, when the state realized that experts were anything but apolitical—such was the case of the fascist managers of LL&P—it sought to balance their political inclinations by courting American experts. With the end of the Second World War, the era of global ideas of development began, embedded in Cold War dynamics. The Peruvian state acquired planning capabilities informed by the Western European experience and received more capital and experts from the United States, as the great power believed that the global conflict could be won through economic development as postulated by Modernization Theory. Finally, and fleetingly, Peru moved closer to the Soviet Union, although the Eastern Bloc was unable to undo decades of linkages between the West and Peru when it came to hydroelectric endeavors. The transnational nature of infrastructural power, hence, should not be ignored.

The aforementioned ideas and expertise were quickly adapted to Peruvian realities, informed by earlier attempts at fostering industrial growth and the local customs that were imprinted upon modernization efforts. Hydroelectric projects tended to be successful because they could be transmitted and communicated to local populations in familiar terms, as experts quickly found out when promoting a rural electrical cooperative or high modernist plants. Global and universal ideas regarding science and development abounded, but they became distinctly Andean as they changed and were changed by local dynamics.

While the pursuit of hydroelectricity highlighted critical divisions in Peru's political class, it remains the best example of the Peruvian state's infrastructural prowess. In the end, however, perhaps this book serves as a cautionary tale, for Peru's—and the Global South's—obsession with infrastructure can be problematic. The construction of infrastructure remains the raison d'être of any state, but twentieth-century Peruvians, with notable exceptions, saw infrastructural development—especially hydroelectricity—as a solution to all the country's ills. While there is no doubt that states need to expand their "infrastructural power," they should consider that this power is always contested and reimagined, certainly when speaking of at least one of its elements, that of physical infrastructures. After all, the Peruvian state has been consistently expanding its infrastructural power in this regard at least since the days of guano. Yet, the Peruvian state is still seen by scholars, and by its very own citizens, as weak, despite its demonstrable and growing technological capacities. The reasons for the weakness of the Peruvian state must be found elsewhere.

Perhaps, then, the problem is one of expectations. To be sure, Peruvian experiences with hydroelectric infrastructures were complex, and while disillusionment abounded regarding the failure to industrialize the country, national hydroelectric endeavors were successful in other ways. First, the state, in alliance with private actors, built hydroelectric plants whose power was readily consumed by citizens, who continue to do so to this day. Second, national and European engineers based in Peru managed to overcome technical obstacles through innovation and ingenuity and produced some excellent science from the periphery. Finally, it offered an avenue through which the Peruvian state could portray its Andean geography in a positive light. But elites would have wished for more. Since independence, the pursuit of infrastructural development by various Peruvian governments has occurred in lieu of addressing the country's social, economic, and political inequalities. Hydroelectricity, elites were sure, would solve all these challenges without the need for structural change. However, the economic development that was to result from electric repetition did not manifest itself. Certainly, Peruvian industry went through a limited expansion during these years, although it is difficult to link this growth to the availability of cheap power. In any case, it remained a small percentage of the national economy, with the export of raw materials, both mineral and agricultural, still the main economic activity. Infrastructure, then, is no substitute for structural change.

It was nevertheless politically expedient to embrace infrastructural politics, as it allowed the state to avoid social and political involvement with the population by building high modernist projects in isolated areas, hoping that they would impact the country's social and economic structures from a distance. Hydroelectricity meant avoiding land reform, avoiding the creation of a more egalitarian society by learning to overcome the cultural differences that divided the country, and, most importantly of all, accepting the Andes as an equal partner in the search for development. By the time a radical military government attempted to do so, it was too late, and in any case, like most military governments, it tried to impose its designs on society rather than seek negotiation and consensus, as was demonstrated by the case of cooperative development in the Mantaro Valley, the only electrification endeavor that was embedded with local social dynamics. Such a disconnect from society—despite being physically connected by the grid—also allowed electricity to be used and abused by various political movements during the twentieth century—from APRA to the Shining Path—who, by capturing infrastructures, could also capture the state's infrastructural power.

Today, Peru's neoliberal governments have gone back into the shadows and show open disdain for any form of economic planning, apart from the construction of infrastructure. Perhaps, as in the past, playing the game of infrastructural politics has proven easier than carrying out far-reaching reforms. Nineteenth-century statesmen may not have had the ability to envision carrying out profound changes in Peru's fragmented society, instead seeking solutions in the construction of railroads. Twentieth-century Peruvian elites understood the need to carry out structural changes but resisted them, seemingly seeking refuge in colossal projects like the construction of hydroelectric plants. It remains to be seen if present-day Peruvians will repeat the mistakes of their predecessors or rise above them by once again reimagining their "traditional" Andean heartlands.

ACKNOWLEDGMENTS

Completing a book is a monumental task, at times feeling as challenging as the very infrastructural projects that this work seeks to study. Much like those projects, this book would have been impossible without the support and guidance of others, in this case a community of transnational scholars who, from the very outset, believed in and nurtured the pages that follow.

This book began as a doctoral dissertation at Stony Brook University. It was there that I found a collegiate group of brilliant scholars from whom I've learned both classic and new historical perspectives that molded my training as a historian. I would like to especially thank Michael Barnhart, Robert Chase, Alix Cooper, Lori Flores, Ned Landsman, Gene Lebovics, Gary Marker, Ian Roxborough, Wolf Schäfer, Shobana Shankar, and Kathleen Wilson. Roxanne Fernandez and Susan Grumet were also instrumental in helping me navigate the SUNY system. As for my fellow graduate students, my relationship with them was characterized by a spirit of collaboration and cooperation that is not always present in all universities. Proposals were exchanged, we read chapters for one another, and words of encouragement were never wanting. I would like to thank Ashley Black, Zinnia Capó Valdivia, Matthew Ford, Carlos Gómez Florentín, Matthew Heidtmann, Matías Hermosilla, Brendan Junkins, Gregory Lella, Kevin Murphy, Liliana Mutu-Blackstone, Emmanuel Pardo, Sergio Pinto-Handler, David Purificato, Leah Savage, Richard Tomzack, Maria-Clara Torres, and David Yee.

In many ways, the origins of this book predate my arrival in the United States. My grandfather, Emilio Romero, was a noted geographer and historian, a well-rounded *pensador*, as he would have been known in Peruvian academia. He was already at an advanced age when I was born and passed away before I could truly know him, yet his legacy was always present, be it in family discussions or in the books that were in my relatives' libraries. As his ideas inevitably found their way into the first chapter of the book, it became clear that this project was an opportunity to connect with him. I hope that, whatever this book's imperfections, it does some honor to the family name.

Other figures in Peru were also highly influential. When I was a master's student at the Pontificia Universidad Católica del Perú, Sinesio López and

Javier Alcalde introduced me to Peruvian political and diplomatic history. I was also fortunate enough to work with Ignacio Basombrío, who would enthusiastically share with me endless anecdotes of Peruvian politics. Sadly, Ignacio passed away before this project reached its culmination. Finally, in 2013, Henry Pease graciously asked me to coauthor a book on the political history of twentieth-century Peru. Henry also passed away over a decade ago, but the lessons I learned from him have accompanied me throughout my career. I am fortunate to have returned to the Pontificia Universidad Católica del Perú to continue to celebrate their legacy.

As this project transitioned from a doctoral dissertation into a book, several other scholars left their mark on it. I would like to mention Sebastien Adins, Rosa Alayza, Cristobal Aljovín, Jorge Aragón, José Carrasco Weston, Omar Coronel, Mercedes Crisostomo, Eduardo Dargent, José Carlos de la Puente Luna, Paulo Drinot, Brenda Elsey, Adrián Lerner, Jorge Lossio, Willie Hiatt, Martín Monsalve, Ricardo Portocarrero, Juan Manuel Rubio, Lizardo Seiner, Teresa Vergara, and Oscar Vidarte. I would like to give special thanks to Javier Puente, who read early versions of this book, and Mark Rice, for reading later drafts more times than he probably cared to. Both Javier and Mark are true friends and have taught me the many tricks of a historian's trade.

Research for this project was financed by various institutions. The Tinker Foundation and the Conference on Latin American History provided early grants, which were essential for establishing the foundations of this study. A Mellon International Dissertation Research Fellowship from the Social Science Research Council allowed me to carry out a year of research both in Peru and the United States, and a Mellon/ACLS Dissertation Completion Fellowship from the American Council of Learned Societies was instrumental in finishing a draft of this book as a dissertation. I would be remiss if I did not thank the staff of the various archives, libraries, and other institutions I visited. Special mention goes to Shiri Alon at the World Bank Group Archives, Julio Nuñez at the Pontificia Universidad Católica del Perú, Eduardo Zapata at the Asociación Electrotécnica Peruana, Axel Koch for improving a key image, and Jay Driskell for some last-minute research assistance at NARA. Many thanks to the staff of the Mantaro Hydroelectric Plant, who took time from their hectic schedule to show a curious historian the workings of that monumental infrastructure. I also express my gratitude to Luis Carlos Arroyo, William Evensen, Tomás Gonzáles, Jaime Guerra, Taras Prytula, Segundo Reymundo, Aníbal Tomecich, and Azi Wolfenson, for taking the time to share past experiences with me. Finally, Neydo Hidalgo at the Museo de la

Electricidad in Lima was, and remains, a close ally. An expert on the field, Neydo opened the doors of the museum for me and generously gave me access to its archives, without which this book would not have been possible.

A small cadre of people truly accompanied this project from beginning to end. Eric Zolov remains an influential figure, as he broadened my understanding of International Relations, helping me to only rely on classic diplomatic perspectives no longer. Brooke Larson changed my very understanding of the Andes. When events seemed to overwhelm me, she always manifested her faith both in me and the project. Finally, I cannot stress enough the influence that Paul Gootenberg has had not only on this book, but on my career. Working with Paul was as challenging as it was rewarding, but not a single page would have been written without his support and advice. His understanding of Peruvian history is unique, and I have benefited immensely from his eclectic intellectual insights, as he pushed me to engage with the history of my country on my own terms, avoiding the often-common fatalistic assumptions regarding Peru's past. Only after working with him for many years can I say that I truly became a historian.

At the University of Nebraska Press, I must thank Bridget Barry, who first expressed interest in publishing my work, and Emily Casillas, who oversaw the project and patiently guided me during the publishing process. I am also thankful to Abigail Kwambamba, project editor, and above all, Stephanie Marshall Ward, for her stoicism in copyediting this book.

Last but certainly not least, I would like to thank loved ones. Jeffrey Green and Rita Nezami became close friends as soon as I arrived in the United States, and without their support I would not have survived for very long. María Cangalaya has been a longtime and loyal friend. The extended Romero family, always a source of inspiration and support. Celeste Arpasi, for her endless patience. My sister María Sol Romero and her husband Carlos Ferraro, for always showing an unbridled enthusiasm for everything I do. My parents, José Romero and Silvia Sommer, for their unquestioning love and faith in me, and, well, for absolutely everything. Finally, I thank Daniela de la Puente, with whom I shared both the excitement and stress of living in New York City, and who has accompanied me through this decade-long journey.

Words will never be enough to thank them all.

NOTES

INTRODUCTION

1. DESCO, *Violencia política en el Perú*, 47.
2. For a very thorough overview of the many concepts dealing with infrastructure and politics, see: Téllez Contreras, "Infrastructural Politics," 31–51.
3. While Peruvian histories of infrastructure are not abundant, they are far from being terra incognita. For a thorough analysis of how elites imagined the linkages between railroads and early industrialization, see Gootenberg, *Imagining Development*. For a strictly economic overview, see Pennano, "Desarrollo regional y ferrocarriles en el Perú," 131–51. Histories of road-building are far more copious. See Contreras and Cueto, "Caminos, ciencia y estado en el Perú," 635–55; Contreras, "La economía del transporte en el Perú," 59–81; Perz and Castillo Hurtado, *The Road to the Land of the Mother of God*; and Rice, "Roads to Progress." A cultural history of aviation can be found in Hiatt, *The Rarified Air of the Modern*. Finally, solid institutional studies regarding the history of electricity in Peru are also available. See Bonfiglio, *Historia de la electricidad en Lima*; Hidalgo, *Tejedores de luz*; and Wolfenson, *El gran desafío*.
4. The modern Peruvian state has long been ignored in this regard, although a valuable ethnography of the state in colonial times can be found in Mumford, *Vertical Empire*.
5. Anand, Gupta, and Appel, *The Promise of Infrastructure*. Another valuable edited volume is Harvey, Jensen, and Morita, *Infrastructures and Social Complexity*. A specific application of these anthropological approaches to the electricity sector can be found in Coleman, *A Moral Technology*.
6. Larkin, "The Politics and Poetics of Infrastructure," 332–34.
7. See once more: Téllez Contreras, "Infrastructural Politics."
8. Mann, "The Autonomous Power of the State," 109–36.
9. Scholars from the region have begun using Mann's ideas to analyze Latin American states through studies of education, taxation, and military recruitment, but rarely actual physical infrastructures. Given that many of these works are comparative in nature—a common trend in the field—they must necessarily simplify the multiplicity of factors that have fostered national infrastructural power. For these valuable works of political theory, see Centeno, *Blood and Debt*; Soifer, *State Building in Latin America*; vom Hau, "State

Infrastructural Power and Nationalism," 334–54; and vom Hau and Biffi, "Mann in the Andes," 191–216. For an analysis of the construction of physical infrastructures and the infrastructural power of the state outside of Latin America, see Naqvi, *Access to Power*. For the relationship between water and power in the region see Hill, "Circuits of State," 13–38.

10. Larson, *Trials of Nation Making*, 150.
11. Hirschman, "Obstacles to Development," 385–93.
12. It is no surprise that many of the new histories of Latin American electrification have centered on its environmental impact on the peoples of the Southern Cone, no doubt attracted by the colossal magnitude of the Itaipu Dam on the Paraguay-Brazil border, and the environmental impact on its people. See: Folch, *Hydropolitics*; Blanc, *Before the Flood*.
13. See Gootenberg, *Imagining Development*. Ideas regarding the mixture of "traditional" Inca practices and "modern" scientific endeavors can also be seen in Gootenberg's *Andean Cocaine*.
14. John Murra spoke of a "vertical archipelago" and highlighted Inca ability to incorporate diverse geographical zones at different altitudinal levels into their empire. See Murra, *El mundo andino*.
15. The direct inspiration for this phrase must inevitably be attributed to the Canadian historian Harold Innis. Innis, *The Fur Trade in Canada*, 397.
16. For the concept of hydraulic empires or civilizations see Wittfogel, *Oriental Despotism*. References to the Incas are found mostly in the first half of the book.
17. Flores Galindo, *In Search of an Inca*.
18. Joseph and Spenser, *In from the Cold*.
19. Contreras and Drinot, "The Great Depression in Peru," 102–28.
20. The few studies that deal with this subject do so from a technocratic viewpoint, and only from the 1980s onward. See Dargent, *Technocracy and Democracy in Latin America*.
21. Rostow, *The Stages of Economic Growth*.
22. For two classic histories on Peru's economy see Sheahan, *Searching for a Better Society*; and Thorp and Bertram, *Peru, 1890–1977*.
23. Evans, *Embedded Autonomy*.
24. Joseph and Spenser, *In from the Cold*.
25. For analyses of the origins of Modernization Theory see Gilman, *Mandarins of the Future*; Latham, *Modernization as Ideology*.
26. For the application of Modernization Theory in the third world, see Engerman et al., *Staging Growth*; Latham, *The Right Kind of Revolution*. The work of Timothy Mitchell is also key, as he argues that, in the Egyptian case, Modernization Theory was used as a postcolonial mechanism of economic control. See: Mitchell, *Rule of Experts*. Finally, Christopher Sneddon has also established the parameters of a "concrete revolution" in the twentieth century, as

he analyzes the role played by the United States when promoting the construction of large dams in the third world in the context of Cold War politics. See: Sneddon, *Concrete Revolution*. Although they do not belong to the Global South, we may speak of some outliers that do not fit comfortably in the West, whose electrification processes bear more similarity to the former than the latter, although Modernization Theory did not guide their actions. Such is the case of countries like the Soviet Union and Spain. See Coopersmith, *The Electrification of Russia*; and Swyngedouw, *Liquid Power*. Finally, given that Modernization Theory reached its high point in the 1960s, one may also use infrastructure as a new avenue to explore what has been dubbed as the "global sixties." See Zolov, "Introduction: Latin America in the Global Sixties."

27. Regarding electricity, the work of Chikowero, "Subalternating Currents," 287–306; Kanduza, "'Let There be Light,'" 39–46; and Tischer, *Light and Power for a Multiracial Nation* show how colonial governments in sub-Saharan Africa used electricity as a tool of racial politics, as well as engaging with political debates that surrounded the public-private nature of the sector.
28. Hughes, *Networks of Power*.
29. Jonnes, *Empires of Light*.
30. McCook, *States of Nature*.
31. Cueto, *Excelencia científica en la periferia* and *Saberes andinos*. Initially, these histories of science were limited to medical practices, but new works have expanded them into other fields. Recent literature has shown how the Peruvian state selectively borrowed local knowledges on the connection among glaciers, lakes, and landslides when it came to managing climate change in the Andes. Likewise, local scientists contributed to furthering technocratic ideals through conservation practices in the aftermath of Peru's nineteenth-century guano boom. This book brings these peripheral histories of science into the realm of electrification. See Carey, *In the Shadow of Melting Glaciers*; Cushman, *Guano and the Opening of the Pacific World*.
32. Pretel, Inkster, and Wendt, "Technology in Latin American History," 1–22.
33. Hausman, Herner, and Wilkins, *Global Electrification*.
34. For the role of experts, see Chastain and Lorek, *Itineraries of Expertise*; and Birn and Necochea López, *Peripheral Nerve*.
35. Scott, *Seeing Like a State*.
36. Scholars have shown how in the construction of roads, engineers experimented with innovative building materials suggested by local populations. See Harvey and Knox, *Roads*.
37. For Rostow's view on the "traditional society" see *The Stages of Economic Growth*, 4–6. For an alternative viewpoint, see Hirschman, "Obstacles to Development," 386–88.

38. A by no means exhaustive list of works on the central highlands would include: Contreras, *Historia económica del Perú central* and *Mineros y campesinos en los Andes*; Flores Galindo, *Los mineros de la Cerro de Pasco*; Long and Roberts, *Miners, Peasants and Entrepreneurs* and *Peasant Cooperation and Capitalist Expansion in Central Peru*; Mallon, *Peasant and Nation*; Manrique, *Mercado interno y region*; Puente, *The Rural State*.
39. Smith, *Uneven Development*.
40. The landmark work remains Cronon, *Nature's Metropolis*. For a similar approach to electrification see Needham, *Power Lines*; and for the development of energy in general, see Jones, *Routes of Power*.
41. Jakle, *City Lights*; Rose, *Cities of Light and Heat*. For a regional history of European electrification, see: Lagendijk, *Electrifying Europe*. There seems to be a shift in this trend, with new studies of urban electrification in the Global South, and rural electrification in the Global North. See: Montaño, *Electrifying Mexico* and Glaser's *Electrifying the Rural American West*.

1. FRAGMENTED POWER

1. The *selva*, or rainforest, was originally named *montaña* in Spanish.
2. Miró Quesada, *Elementos de geografía científica del Perú*, 18.
3. Pratt, *Imperial Eyes*. See also Poole, *Vision, Race, and Modernity*.
4. Anand, Gupta, and Appel, *The Promise of Infrastructure*.
5. Harrison-Moore and Sandwell, *In a New Light*; as well as Gooday, *Domesticating Electricity*.
6. Mann, "The Autonomous Power of the State,"117.
7. The phrase can be attributed to Jorge Basadre, see: Basadre, *Historia de la República del Perú*. For key works on the history of the guano, see Bonilla, *Guano y burguesía*, Gootenberg, *Between Silver and Guano*, and Cushman, *Guano and the Opening of the Pacific World*.

For a recent approach see Sobrevilla Perea, "¿Qué tan falaz fue la prosperidad?" 95–142.

8. For a general overview of energy sources during this period, see: Lizarme Villcas and Amaya Núñez, "Energías del Centenario," 17–29.
9. Orlove, "Putting Race in Its Place." For von Humboldt in Peru see Pratt, *Imperial Eyes*. For an overview of the role of American scholars on the continent in the early twentieth century—including Isaiah Bowman in Peru—see: Salvatore, *Disciplinary Conquest*.
10. Mariano Felipe Paz Soldán also published the enormously popular *Diccionario Geográfico Estadístico del Perú*, a work that did much to create a sense of order out of Peru's "chaotic" nineteenth-century geography.
11. Orlove, "Putting Race in Its Place," 322–26.

12. On the link between the Sociedad Geográfica de Lima and the Peruvian state, see Cueto, "Apogeo y crisis de la Sociedad Geográfica de Lima," 36. See also Palacios Rodríguez, *La Sociedad Geográfica de Lima* and López-Ocón, "La Sociedad Geográfica de Lima."
13. Carranza, *Boletín de la Sociedad Geográfica de Lima*, vol. 1 (1892): 1–2. For a thorough analysis of Carranza's view of the Andean environment, see: Palomo, "Embracing the Cordillera," 1–20.
14. For a brief overview of the European experience with White Coal see: Landry, "Water as 'White Coal,'" 7–11.
15. Stiglich, *Geografía comentada del Perú*, 85
16. Miró Quesada, *Elementos de geografía científica del Perú*, 221.
17. The region on the eastern slopes of the Andes is also known as *ceja de selva*; no such term exists for the western slopes.
18. Ricardo García Rosell, "La irrigación de la costa del Perú," 125–26. Although one of the earliest, García Rosell was not the first to champion this, as one of the first mentions of the use of water for power can be found in Manuel Pardo's 1862 essay "Estudios sobre Jauja," in which he analyzed the possibilities for the creation of a textile industry in the midst of the guano boom: "cheaper wages and foodstuffs, primary materials, ready coals and better yet—powerful waterfalls—why can't Peru establish factories for coarse cloth, rough cottons, and linens, or those for ordinary pottery, for leather goods, and a host of potassium and chemical products." Quoted in Gootenberg, *Imagining Development*, 84.
19. García Rosell, "La irrigación de la costa del Perú," 127.
20. Garland, "Importancia de los ríos peruanos," 101.
21. Delgado, "Memoria correspondiente al año 1905," 14.
22. Delgado, "Memoria correspondiente al año 1905," 15.
23. Coronel Zegarra, "El ferrocarril de Paita al Marañón," 449.
24. Coronel Zegarra, "El ferrocarril de Paita al Marañón," 451.
25. Alfonso Pezet, "Estudio de la colonización del Perú, bajo el punto de vista práctico," 131.
26. For an overview of the Generación del 900, see Planas Silva, *El 900*.
27. García Calderón, *El Perú contemporáneo*, 4.
28. García Calderón, *El Perú contemporáneo*, 5.
29. This is a recurring theme in Republican Peruvian history, See: Méndez, "Incas Sí, Indios No."
30. García Calderón, *El Perú contemporáneo*, 21.
31. García Calderón, *América Latina y el Perú del novecientos*, 82.
32. García Calderón, *El Perú contemporáneo*, 123.
33. García Calderón, *El Perú contemporáneo*, 168.
34. García Calderón, *El Perú contemporáneo*, 182.

35. de la Riva-Agüero y Osma, *Paisajes peruanos*, 32. By the "romantics," de la Riva-Agüero was referring to the eighteenth- and nineteenth-century naturalists and writers who had explored the sierra.
36. de la Riva-Agüero y Osma, *Paisajes peruanos*, 32.
37. de la Riva-Agüero y Osma, *Paisajes peruanos*, 187.
38. Enock, *The Andes and the Amazon*, 1.
39. Enock, *The Andes and the Amazon*, 2.
40. Enock, *The Andes and the Amazon*, 162.
41. Enock, *The Andes and the Amazon*, 170.
42. Enock, *The Andes and the Amazon*, 170.
43. "Hydraulic Power from Andean Waters," *Monthly Consular Trade Reports*, no. 297 (June 1905): 200.
44. Guarini, *El porvenir de la industria eléctrica*, 30. For more details on the life of Emilio Guarini, as well as his failed project of establishing a National School of Electricity, see: Guizado Mercado and Ragas, "The National School of Electricity and Early Technical Education in Peru," article 15.
45. Guarini, *El porvenir de la industria eléctrica*, 100.
46. Guarini, *El porvenir de la industria eléctrica*, 101–2.
47. A similar project would be proposed some decades later, on the Bolivian side of the border, by the engineer Arthur Posnansky, although he presented it as a regional venture that could result in the "economic independence" of Argentina, Bolivia, Chile and Peru. See Posnansky, *A los pueblos de Argentina, Bolivia, Chile y Perú*.
48. "Un proyecto gigantesco: Una entrevista con Mr. Guarini," *El Comercio*, January 4, 1906.
49. "Proyectos nuevos," *El Comercio*, January 20, 1906.
50. P. Strange, "Two Electrical Periodicals: The Electrician and The Electrical Review 1880–1890," IEE *proceedings. Part A, Physical Science, Measurement and Instrumentation, Management and Education*, reviews, vol. 132, 8, (1985): 574.
51. Pepper, "Electricity in Peru," 319.
52. Guarini, "La moderna enseñanza técnica y el porvenir de la enseñanza obrera en el Perú," 212. For a general overview of technical education during this period, see: Guizado Mercado, "De la palabra a la mano," 71–84; and Lizarme Villcas, "Education and Progress."
53. "The School of Engineers in Lima: New Electrical Installation," *Peru Today* 4, no. 3 (June 1912): 137.
54. "The Electrical Department of the National School of Arts and Crafts (Lately Reorganized by an American Graduate)," *Peru Today*, 4, no. 8 (November 1912): 430–31.

55. Nineteenth-century intellectuals such as Luis Esteves also believed that if freed from the oppression of Peru's elites and given scientific tools, the Indians could become a "civilizing" force. For the economic ideas of Luis Esteves, see Gootenberg, *Imagining Development*, chapter 6.
56. *La obra de los ingenieros en el progreso del Perú*, 9.
57. Guarini, *El porvenir de la industria eléctrica*, 547.
58. Tizón y Bueno, *La reorganización del Ministerio de Fomento*, 11–12.
59. Tizón y Bueno, *La reorganización del Ministerio de Fomento*, 31.
60. Chiozza Money, "Why Should Scotland Lose Population?" 146.
61. Smith, "Electrical Goods in Peru and Ecuador," 21.
62. "Electric Installations in Peru," *West Coast Leader*, Centennial Number, Peruvian Yearbook, October 1921, 32.
63. Smith, "Electrical Goods in Peru and Ecuador," 21.
64. "Mining Camps on the Central: Cerro de Pasco," *West Coast Leader*, July 6, 1916, 9.
65. Water Code of 1902, https://leyes.congreso.gob.pe/Documentos/LeyesXIX/1902007.pdf.
66. Mining Code of 1900, Biblioteca del Congreso de la República, https://www.leyes.congreso.gob.pe/Documentos/LeyesXIX/1900065.pdf.
67. Ministerio de Fomento, *Padrón de fuerza motriz hidráulica*. 89–91.
68. "Decree of March 2, 1937, re: Government grants of water and land concessions," March 8, 1937. RG 59, 1930–39 Central Decimal File, file 823.6461/, box 5733, NARA.
69. The Ministry of Fomento and Public Works is usually translated as the Ministry of Development, although the word fails to communicate the true purpose of the institution, which was to *promote* rather than develop, as the modern idea of development was still evolving. For one of the few histories written about this ministry see: Quiñones Tinoco, *Construir y modernizar*.
70. Mann, "The Autonomous Power of the State," 117.
71. Tizón y Bueno, *La reorganización del Ministerio de Fomento*, 51.
72. Portocarrero C., *Contribución al estudio de los recursos hidráulicos para fuerza motriz en el Perú*, 44–47.
73. Lara, *Historia del Ministerio de Fomento*, 256–60.
74. Lara, *Historia del Ministerio de Fomento*, 174–75.
75. "Carta de Acidalio Ortiz Silva al director de Obras Públicas," May 28, 1928. Expedientes 64, 65, box 181, MTC-SDE; "Carta de Acidalio Ortiz Silva al director de Obras Públicas," November 9, 1928. Expedientes 64, 65, box 181, MTC-SDE.
76. "Carta de Acidalio Ortiz Silva al director de Obras Públicas," May 28, 1928. Expedientes 64, 65, box 181, MTC-SDE; "Carta de Acidalio Ortiz Silva al director de Obras Públicas," November 4, 1928. Expedientes 64, 65, box 181, MTC-SDE.

77. "Carta de Víctor Cáceres al director de Obras Públicas," April 28, 1932. Expedientes 64, 65, box 181, MTC-SDE.
78. "Servicio deficiente de alumbrado eléctrico en localidad de Lircay," *La Voz de Huancayo*, June 24, 1960.
79. Smith, "Electrical Goods in Peru and Ecuador," 21.
80. Smith, "Electrical Goods in Peru and Ecuador," 27.
81. For the modernization of Lima under Leguía, see: Paredes Hernández, "Arquitectura y discurso simbólico en el Oncenio de Leguía," 33–47; and Martuccelli Casanova, "Lima, capital de la Patria Nueva."
82. Alberto Arca Parró, "La ciudad capital de la República y el Censo Nacional de 1940," *Estadística Peruana*, vol. 1 (1945): 28. The poetry of phrase can be attributed to Antúnez de Mayolo, *La génesis de los servicios eléctricos de Lima*, 1.
83. Muñoz Cabrejo, *Diversiones públicas en Lima 1890–1920*, 54.
84. Antúnez de Mayolo, *La génesis de los servicios eléctricos de Lima*, 2.
85. "The Centennial Feasts: Some Impressions on the Illuminations," *West Coast Leader*, Centennial Number, Peruvian Yearbook, October 1921, 19–20.
86. "The Centennial Feasts: Some Impressions on the Illuminations," *West Coast Leader*, 19–20.
87. For social reactions to the arrival of electricity, see Schivelbusch, *Disenchanted Night*.
88. José Gálvez was a native of Tarma, Junín, but moved to Lima when he was nine years old. He was the son of an engineer.
89. Gálvez, *Obras Completas*, vol. 2, 29.
90. Gálvez, *Obras Completas*, vol. 4, 131.
91. Arana, *La electricidad en el hogar*, 53.
92. Arana, *La electricidad en el hogar*, 55–56.
93. Harrison-Moore and Sandwell, *In a New Light*; and Gooday, *Domesticating Electricity*.
94. United States Bureau of Foreign and Domestic Commerce, *Electrical Development*, 12.
95. United States Bureau of Foreign and Domestic Commerce, *Electrical Development*, 12.
96. Antúnez de Mayolo, *La génesis de los servicios eléctricos de Lima*, 42.
97. Smith, "Electrical Goods in Peru and Ecuador," 28–29.
98. Mariátegui, *Siete ensayos*, 172.
99. Mariátegui, *Siete ensayos*, 178.
100. Mariátegui, *Siete ensayos*, 21.
101. Mariátegui, *Siete ensayos*, 177.
102. "El plan quinquenal para la industria del estado," 8.

103. One may question Mariátegui's analysis in this regard. London was not Manchester, Berlin was no Ruhr Valley, and New York—which was not a capital—was not Pittsburgh. Financial, political, and military considerations had proved just as important as industrial development in determining the political dominance of certain European capitals and cities in general.
104. Mariátegui, *Siete ensayos*, 178.
105. Romero Padilla, *Geografía económica del Perú*, 3–4.
106. By the 1930s, the Sociedad Geográfica de Lima had entered a period of decline. Many of its tasks were taken over by the army and the newly created Servicio Geográfico del Ejército. Furthermore, Peruvian researchers determined that scientific research required greater specialization, and, despite its name, the Sociedad Geográfica de Lima incorporated a number of diverse voices. As Marcos Cueto points out, it was precisely the work of Emilio Romero and others—such as Javier Pulgar Vidal—that breathed new life into the discipline of geography, although this did not halt the declining importance of the institution. See Cueto, "Apogeo y crisis," 44–45.
107. Romero Padilla, *Geografía económica del Perú*, 7.
108. Proposals for decentralization became commonplace in the 1930s after the centralism of Leguía's government. See Planas, *La descentralización en el Perú Republicano*. Aside from Romero Padilla, Javier Pulgar Vidal also proposed rearranging the country's geography by creating eight natural regions, using what he termed "traditional geographical knowledges" of Peru's Indians. See Pulgar Vidal, *Geografía del Perú*, 302.
109. Romero Padilla, *El descentralismo*, 10.
110. Romero Padilla, *El descentralismo*, 31–41.
111. Romero Padilla, *Geografía económica*, 395–96.
112. Romero Padilla, *Geografía económica*, 397.
113. Pareja Paz Soldán, *Geografía del Perú: curso universitario*, 3. Romero Padilla, while an *indigenista*, agreed, stating that "creating antagonism between whites and Indians is infantile." Pareja Paz Soldán duly quoted him in his book.
114. Pareja Paz Soldán, *Geografía del Perú: curso universitario*, 4.
115. Pareja Paz Soldán, *Geografía del Perú: manual*, 107.
116. Pareja Paz Soldán, *Geografía del Perú: manual*, 109.
117. Pareja Paz Soldán, *Geografía del Perú: curso universitario*, 164.
118. Pareja Paz Soldán, *Geografía del Perú: curso universitario*, 162.
119. Pareja Paz Soldán, *Geografía del Perú*, vol. 1, 228.
120. Pareja Paz Soldán, *Curso de geografía del Perú*, 197.
121. Pareja Paz Soldán, *Curso de geografía del Perú*, 197.
122. Basadre, *Perú: problema y posibilidad*, 238.
123. Mann, "The Autonomous Power of the State," 118.
124. Mariátegui, *Siete ensayos*, 180.

2. VERTICAL LIMITS

1. "La actualidad," *El Comercio*, May 2, 1908.
2. "La actualidad," *El Comercio*, May 2, 1908.
3. Winner, "Do Artifacts Have Politics?" 189–204.
4. Drinot, *The Allure of Labor*.
5. Joseph, LeGrand, and Salvatore, *Close Encounters of Empire*.
6. Leonard and Bratzel, *Latin America during World War II*.
7. Hausman, Herner, and Wilkins, *Global Electrification*.
8. Cueto, *Excelencia científica en la periferia*.
9. Such a view problematizes ideas present in Scott, *Seeing Like a State*.
10. Bonfiglio, *Los italianos en la sociedad peruana*, 201.
11. Although commonly referred to as one, the Huatica itself was not, strictly speaking, a river. It had originally been a pre-Colombian irrigation canal that used the waters of the Rimac River, which in turn had been modified by the Spanish to provide water to the city.
12. Empresas Eléctricas Asociadas, *60 años*, 15.
13. Basadre, *Historia de la República del Perú*, vol. 11, 168.
14. For an overview of the Prado's business concerns, see Portocarrero, *El Imperio Prado*.
15. Piaggio would also eventually become a key figure in the early development of Peru's oil industry.
16. Antúnez de Mayolo, *La génesis de los servicios eléctricos de Lima*, 3–4.
17. "El trust eléctrico de Lima," *El Comercio*, July 15, 1906. According to Pastor Campos, the amount would have roughly equaled 7,000,000 U.S. dollars at the time. However, estimates by Dargent Chamot would place the figure significantly lower, at 4,000,000 U.S. dollars. In any case, either number was indeed unprecedented. Pastor Campos, "Peru: Monetary and Exchange Rate Policies"; Dargent Chamot, *El billete en el Perú*.
18. Empresas Eléctricas Asociadas, *60 años*, 16. For an overview of these financial groups see: Monsalve Zanatti, "Evolution of the Peruvian Large Family Business."
19. Roberts, *Schroders*, 139.
20. "El porqué de la denominación Lima Light and Power," *Kilowatito*, no. 38, October 1970; Hidalgo, *La profesión civilizadora*, 81.
21. Empresas Eléctricas Asociadas, *60 años*, 16–18.
22. Pepper, "Electricity in Peru," 320.
23. Empresas Eléctricas Asociadas, *Souvenir of the Empresas Eléctricas Asociadas*, v.
24. Pepper, "Electricity in Peru," 320.
25. Empresas Eléctricas Asociadas, *60 años*, 18.
26. Pepper, "Electricity in Peru," 320.

27. "High-Tension Energy Transmission in Peru," 223.
28. Guarini, *El porvenir de la industria eléctrica en el Perú*, 741–42.
29. "La Solución de la huelga," *El Comercio*, December 23, 1906. For a thorough labor history of the electricity sector see: Castillo Paulino, *Seguridad y salud en el trabajo en la industria eléctrica.*
30. "La huelga," *El Comercio*, September 11, 1912.
31. Zitor, *Historia de las principales huelgas y paros obreros habidos en el Perú*, 23.
32. Rowe, *Early Effects of the War*, 31.
33. "New Electric Concessions in Paraguay," *Daily Consular and Trade Reports*, issues 1–75 (1911): 248.
34. "Recordando a Juan Carosio," *Kilowatito*, no. 33, July 1969.
35. Bardella, *Un siglo en la vida económica del Perú*, 154–55.
36. Bonfiglio, *Historia de la electricidad en Lima*, 36–37.
37. "Recordando al Dr. Bianchini," *Kilowatito*, no. 29, July 1968.
38. Law no. 4510, March 24, 1922, Biblioteca del Congreso de la República, https://www.leyes.congreso.gob.pe/Documentos/Leyes/04510.pdf.
39. "Recordando a Juan Carosio," *Kilowatito*, no. 33, July 1969.
40. *La obra de los ingenieros en el progreso del Perú*, vol. 4, 8–10.
41. Portocarrero S. and Camacho S., "Impulsos moralizadores," 37, 58.
42. Empresas Eléctricas Asociadas, *Los servicios eléctricos en el Callao*, 8–11.
43 "El Hombre de la Calle conmina terminantemente y de una vez por todas, a la Eléctricas Asociadas para que terminen sus líos con el público," *El Hombre de la Calle*, October 25, 1930.
44. "A propósito del informe emitido por el señor Ingeniero Inspector de los Servicios Eléctricos de la Municipalidad de Lima y la contestación diáfana, rotunda y clara de las Empresas Eléctricas," *La Revista Semanal*, October 23, 1930.
45. "El Hombre de la Calle conmina terminantemente y de una vez por todas, a la Eléctricas Asociadas para que terminen sus líos con el público," *El Hombre de la Calle*, October 25, 1930.
46. Mariátegui, *Ideología y política*, 150.
47. Ricardo Martínez de la Torre, "La teoría del crecimiento de la miseria aplicada a nuestra realidad," *Amauta*, no. 27 (November–December 1929): 73.
48. "Las EE. EE. AA. despiden violentamente a tres trabajadores de las centrales de luz," *El Comercio*, October 17, 1931.
49. "Lima Light Power and Tramways Company," *El Comercio*, October 17, 1931.
50. "El Sindicato Central de Generación Chosica, Yanacoto y Santa Rosa y Similares refuta a las EE. EE. AA," *El Comercio*, October 19, 1931.
51. Letter from Fred Morris Dearing to the Secretary of State, October 22, 1931. RG 59, 1930–39 Central Decimal File, file 823.504/29, box 5717, NARA.
52. "Actitud resuelta de los trabajadores," *La Tribuna*, October 16, 1931.

53. "Voto de aplauso a La Tribuna," *La Tribuna*, October 20, 1931.
54. Murillo Garaycochea, *Historia del* APRA, 127.
55. Chanduví, *El* APRA *por dentro*, 48.
56. Martínez de la Torre, *Apuntes para una interpretación marxista*, vol. 3, 489.
57. Martínez de la Torre, *Apuntes para una interpretación marxista*, vol. 3, 487.
58. For the Kemmerer Mission see Drake, *The Money Doctor in the Andes*.
59. Empresas Eléctricas Asociadas, *Memoria del directorio* 1933, 6.
60. Martínez de la Torre, *Apuntes para una interpretación marxista*, vol. 3, 485.
61. "Los problemas monetarios y tarifarios de los servicios eléctricos y tranviarios en Lima y alrededores," *El Comercio*, April 3, 1934.
62. "Dictamen de la comisión consultiva presidida por el Dr. Fernando Gazzani," *El Comercio*, April 8, 1934; "Dictamen de la comisión consultiva presidida por el Dr. Fernando Gazzani," *El Comercio*, April 10, 1934.
63. "Dictamen de la comisión consultiva presidida por el Dr. Fernando Gazzani," *El Comercio*, April 9, 1934.
64. "El conflicto entre las empresas eléctricas y sus asociados," *El Comercio*, April. 22, 1934.
65. Letter from Fred Morris Dearing to the Secretary of State, January 29, 1935. RG 59, 1930–39 Central Decimal File, file 823.00/1140, box 5698, NARA.
66. The union of tramway workers continued its radicalism after the split, stating that the union line responded to the fact that "in all countries where the capitalism system exists, we can see the division of society into two classes with opposing interests: on the one hand the exploited proletariat, on the other the exploiting bourgeoise." See Federación de Motoristas, *Pactos colectivos*, 3.
67. Buse, *Callahuanca*, 7.
68. Bonfiglio, *Historia de la electricidad en Lima*, 109–10.
69. Graubünden State Archives, *Naturalizations 1801–1960*, 172–73, https://www.gr.ch/DE/institutionen/verwaltung/ekud/afk/sag/dienstleistungen/bestaende/kantonalesarchiv/Documents/QR_1-3a_STAR.pdf.
70. Pablo Boner, "El planeamiento de las obras eléctricas." Boner's views on the electricity sector can also be found in Boner and Donizetti, *Los servicios eléctricos*.
71. Boner, "El planeamiento de las obras eléctricas," 10.
72. Buse, *Callahuanca*, 8.
73. Buse, *Callahuanca*, 8.
74. "Una etapa importante de Pablo Boner," *Kilowatito*, no. 14, July 1964; Bonfiglio, *Historia de la electricidad en Lima*, 110.
75. Turner, *Las lagunas de Huarochirí*, 12; Stubbs, *Registro histórico de Lurigancho y Chosica*, 274.
76. Turner, *Las lagunas de Huarochirí*, 12.

77. "Wealthy South American Student Gets Experience in Canadian Mines," *Ohio State Lantern*, October 7, 1930.
78. Ludeña, "Prospecting in the Peruvian Andes," 26.
79. Ludeña, "Prospecting in the Peruvian Andes," 26.
80. Ministerio de Fomento, *Boletín oficial de la Dirección de Minas y Petróleo*, vol. 21–22, no. 70–71, (1943): 436.
81. "Una etapa importante de Pablo Boner," *Kilowatito*, no. 14, July 1964.
82. Buse, *Callahuanca*, 8.
83. For current anthropological perspectives on LL&P activities in the area, see: Hommes and Boelens, "From Natural Flow to 'Working River,'" 85–95.
84. Lasarte, *Obras de represamiento en las lagunas de Huarochirí*, 5.
85. Adams, "Caudal, procedencia y distribución de aguas," 12.
86. "Hoy será inaugurada la central Juan Carosio," *El Comercio*, May 7, 1938.
87. Buse, *Callahuanca*, 11.
88. Ramírez Villacorta, "La penetración capitalista en una comunidad campesina," 51.
89. "Callahuanca," *El Comercio*, April 3, 1954.
90. Buse, *Callahuanca*, 12.
91. "La nueva central Juan Carosio en el Rio Santa Eulalia," *La Prensa*, May 7, 1938.
92. "Callahuanca," *El Comercio*, April 3, 1954.
93. "Hoy será inaugurada la central Juan Carosio," *El Comercio*, May 7, 1938.
94. "Juan Carosio, A Big Modern Hydro-Electric Power Station in Peru," 227.
95. "La Inauguración de la central hidroeléctrica Juan Carosio," *El Comercio*, May 8, 1938.
96. "Ayer se inauguró la nueva central hidroeléctrica Juan Carosio en el valle de Sta. Eulalia," *El Comercio*, May 8, 1938.
97. Boveri, *Juan Carosio, 1876–1959*, 10.
98. For Benavides's policies on labor see Drinot, *The Allure of Labor*.
99. "Ayer se inauguró la nueva central hidroeléctrica Juan Carosio en el valle de Sta. Eulalia," *El Comercio*, May 8, 1938.
100. Whitaker, *Americas to the South*, 18.
101. Letter from Lawrence Steinherdt to the Secretary of State, October 9, 1937. RG 59, 1930–39 Central Decimal File, file 723.65/8, box 4149, NARA.
102. Beals, *The Coming Struggle for Latin America*, 102.
103. Boveri, *Juan Carosio, 1876–1959*, 6–7.
104. Letter from Henry Norweb to the Secretary of State, July 16, 1940. RG 38, Office of Naval Intelligence Monograph Files, Peru, box 58, NARA.
105. Ciccarelli, "Fascist Propaganda and the Italian Community," 367. See also, by the same author, "Fascism and Politics in Peru," 405–32.
106. The episode is narrated in Ciccarelli, "La crisis etíope y la liga de naciones en la prensa peruana," 35–48. For an overview of fascism in Peruvian society,

including the ideas of Carlos Miró Quesada, see: López Soria, *El pensamiento fascista en el Perú*.

107. Ciccarelli, "Fascist Propaganda and the Italian Community," 367.
108. De Vivero, *Avance del imperialismo fascista*, 13.
109. "Conferencia del Señor M. Seoane, sustentada el sábado," *La Prensa*, April 13, 1931.
110. De Vivero, *Avance del imperialismo fascista*, 14. It should be noted that Santa Rosa—omitted in the pamphlet—was a thermal plant that consumed oil.
111. Letter from J. Edgar Hoover to the Assistant Secretary of State, July 11, 1942. RG 59, 1940–1944 Central Decimal File, file 823.00 PRADO/276, box 4349, NARA.
112. Memorandum "Engineer Jones, T.V.A.," November 8, 1941. RG 59, 1940–1944 Central Decimal File, file 823.6463/21, box 4371, NARA.
113. Memorandum "H. A. Brassert and Company," April 15, 1942. RG 59, 1940–1944 Central Decimal File, file 823.6511/30, box 4372, NARA.
114. "Scientific Revolt in Peru May Wreck Plans of Dictators," *St. Louis Daily Globe Democrat* (St. Louis MI), April 5, 1941.
115. Letter from the Ministry of Fomento to the TVA, March 11, 1944. RG 59, 1940–1944 Central Decimal File, file 823.6463/36, box 4371, NARA.
116. Letter from George Butler to the Secretary of State, April 25, 1941. RG 59, file 823.4 engineers society/19, box 4361, NARA.
117. Letter from Jefferson Patterson to the Secretary of State, September 28, 1942. RG 59, file 823.461 /17, box 4363, NARA.
118. Letter from Louis G. Dreyfus to the Secretary of State, December 19, 1939. RG 59, 1940–1944 Central Decimal File, file 823.43-ITALIAN-PERUVIAN CULTURAL INSTITUTE/I, box 4362, NARA.
119. de la Riva-Agüero y Osma, *Obras completas de José de la Riva-Agüero*, 604–5.
120. "Prado Begins Tour of U.S. Arms Centers," *Washington Post*, May 12, 1942.
121. "Prado Praises U.S. Spirit in Abell Interview," *Washington Times Herald* (Washington IN), May 12, 1942.
122. Armando Bueno Ortiz, "Algunos aspectos geopolíticos del Perú en la defensa nacional," 1547. Regarding fears of a Japanese invasion, see: Ulloa y Sotomayor, *Posición internacional del Perú*, 331–63.
123. Empresas Eléctricas Asociadas, *Memoria del directorio: Ejercicio 1939*.
124. To avoid being blacklisted, Italian investors renamed it Banco de Crédito.
125. Graubünden State Archives, *Naturalizations 1801–1960*, 173, https://www.gr.ch/DE/institutionen/verwaltung/ekud/afk/sag/dienstleistungen/bestaende/kantonalesarchiv/Documents/QR_1-3a_STAR.pdf.
126. Bonfiglio, *Historia de la electricidad en Lima*, 111.
127. "Addio, Dottore Bianchini," *Presente*, no. 19, April 13, 1957, 27.
128. Empresas Eléctricas Asociadas—HIDRANDINA, *Electricidad para la Gran Lima*.

129. "Juan Carosio-Moyopampa Hydroelectric Station in Peru," 405.
130. Barrera, "Líneas de transmisión a 64 KV Moyopampa-Lima," 17.
131. Bonfiglio, *Historia de la electricidad en Lima*, 53; "Han sido tomadas algunas medidas para aliviar el problema de energía eléctrica," *La Prensa*, June 8, 1949.
132. "Huampaní, 30,000 KW," *El Comercio*, March 25, 1960.
133. Empresas Eléctricas Asociadas, *60 años*, 41
134. "Después de prestar 54 años de servicios, la central de Yanacoto entrará en receso," *El Comercio*, March 29, 1960.

3. ELECTRIFICATION FROM ABOVE

1. Natella, "Enrique Solari Swayne and Collacocha," 39. For a recent edition of this play, see Solari Swayne, *Collacocha*.
2. While a thorough study of Peru's industrialists during this period is still needed, a general Marxist approximation can be found in Durand, *La decada frustrada*.
3. Evans, *Embedded Autonomy*.
4. Scott, *Seeing Like a State*.
5. Such a divide is also clear in the American case. See: Needham, *Power Lines*.
6. Hausman, Herner, and Wilkins, *Global Electrification*.
7. For Modernization Theory in the rest of the world see: Engerman et al., *Staging Growth*.
8. For a political overview of that tumultuous period, see: Haworth, "Peru," 170–89.
9. Peru President, *A Program of Action*, 8.
10. For an overview of Odría's economic policies, see Thorp and Bertram, *Peru, 1890–1977*. For a celebratory overview of Odría's government see Guerra, *Manuel A. Odría*.
11. Ambassador's Impressions, Memorandum of Conversation, July 12, 1951. RG 59, 1950–54 Central Decimal File, file 823.00/7–1251, box 4591, NARA.
12. For the role of the National Agrarian Society in Odría's seizure of power, see Portocarrero, *De Bustamante a Odría*. For the Klein mission see Kofas, *Foreign Debt and Underdevelopment*. While it is true that Odría was ideologically opposed to the idea of development, Thorp and Bertram have convincingly argued that Peru's industrialization policy came to halt in 1945, during the government of José Luis Bustamante y Rivero, due to political conflict and a lack of a coherent overall economic policy. Thorp and Bertram, *Peru 1890–1977*, 187–88.
13. Wolfenson, *El gran desafío*, 103–109.
14. Empresas Eléctricas Asociadas, *Memoria del directorio: Ejercicio 1948*, 6–7
15. "La S.N. de I. transmite al Sr. Ministro de Fomento su inquietud y preocupación ante la perspectiva de una posible interrupción de los servicios públicos de energía eléctrica," *Industria Peruana* 16, no. 10 (October 1946): 566.

16. Empresas Eléctricas Asociadas, *Memoria del directorio: Ejercicio 1950*, 6–7
17. Empresas Eléctricas Asociadas—HIDRANDINA, *Energía eléctrica para la gran Lima*, 2.
18. "Las barriadas clandestinas," *El Comercio*, October 6, 1955.
19. One of the earliest studies on Lima's barriadas was published three years after Odría's departure. It showed that out of the then existing fifty-six barriadas, thirty already had access to electricity. Matos Mar, *Migration and Urbanization*, 16. For Odría's general policy toward the barriadas see Collier, *Squatters and Oligarchs*.
20. Fritz Vallenas, "La Ley de la Industria Eléctrica y sus consecuencias."
21. Cavers and Nelson, *Electric Power Regulation in Latin America*, 16–19.
22. Ministerio de Fomento y Obras Públicas, *Ley de industria eléctrica No. 12378*.
23. Manuel A. Odría, Speech to Congress, July 28, 1955, https://www.congreso.gob.pe/participacion/museo/congreso/mensajes/mensaje_nacion_congreso_27_julio_1955.
24. Thorp and Bertram also point out that the end of this export boom pushed Odría to rethink his position regarding industrialization. Thorp and Bertram, *Peru 1890–1977*, 261.
25. "Electric Industry Law Promulgated in Peru." Foreign Service Dispatch, August 15, 1955. RG 59, 1955–1959 Central Decimal File, file 823.2614/8-1555, box 4268, NARA.
26. Empresas Eléctricas Asociadas, *Memoria del directorio: Ejercicio 1955*, 5–6.
27. "Ley de industria eléctrica," *Kilowatito*, no. 5, June 1961.
28. Cavers and Nelson, *Electric Power Regulation in Latin America*, 130.
29. Boletín de la Asociación de Empresarios Eléctricos, *Electricidad en el Perú*, no. 5, November 1959.
30. Fritz Vallenas, *La ley peruana de electricidad*, 5.
31. There is also the matter of the amount of electricity generated. The Itaipu Dam, situated on the Paraguay-Brazil border, produces 14 GW, most of which is exported to Brazil. In the case of Peru, the most ambitious project would reach 1 GW.
32. There are no works that specifically deal with Manuel Prado's second government. A general overview can be found in Klarén, *Peru*.
33. The Civilista Party was Peru's first political party. Founded in 1871, by the time of the Aristocratic Republic it represented the interests of an established oligarchy and adhered to a program of economic liberalism. Civilistas, however, were willing to promote infrastructural projects—such as railroads in the nineteenth century—with the hopes of promoting industrialization and "civilizing" Peru's Indigenous populations. For nineteenth-century civilista views on development, see: Gootenberg, "Order(s) and Progress in Develop-

mental Discourse," 111–35; Gootenberg, *Imagining Development*; and Bonilla, *Guano y burguesía en el Perú*.

34. Gobierno del Perú. *Panorama de la electrificación*, 45.
35. Gobierno del Perú. *Panorama de la electrificación*, 43–45.
36. For Gerschenkron's views on the link between technology and economic development, see: Gerschenkron, *Economic Backwardness in Historical Perspective*. For the Peruvian case see Gootenberg, "Hijos of Dr. Gerschenkron," 57–80.
37. Gobierno del Perú, *Panorama de la electrificación*, 43
38. Asociación Electrotécnica Peruana, *Libro de oro*, 49.
39. For the concept of embedded autonomy, see Evans, *Embedded Autonomy*.
40. "Discurso del Ing. Jorge Grieve en el XV Aniversario de la A.E.P," 28.
41. Rodríguez Valencia and Seiner Lizárraga, *Juan Alberto Grieve Becerra y Ricardo Tizón y Bueno*, 34.
42. Despite their enmity with Odría, like many apristas, Grieve would eventually join his political party to form the UNO-APRA coalition and became the party's mayoral candidate in the 1966 elections.
43. Jorge Grieve, "Plan de electrificación nacional," 21.
44. Grieve, "Plan de electrificación nacional," 23–24.
45. "Plan de electrificación—Disposiciones Oficiales."
46. "Discurso del Ing. Jorge Grieve en el XV Aniversario de la A.E.P," 29.
47. Grieve, "Plan de electrificación nacional," 21.
48. Grieve, "Plan de electrificación nacional," 21.
49. Hager, *Hydraulicians in Europe 1800–2000*, 988.
50. Électricité de France, *Plan de electrificación nacional*, v.
51. "Conversatorio sobre el plan nacional de electrificación," 27.
52. Although both engineers trained in the West, the input of Antúnez de Mayolo and Boner should be noted. While they do not represent "traditional knowledge" in the Indigenous sense of the word, they certainly represent "local knowledge," as they were familiar with the Peruvian electric landscape, informing the action of the French mission. Once more, see Cueto, *Excelencia científica en la periferia* and *Saberes andinos*.
53. "Conversatorio sobre el plan nacional de electrificación," 27.
54. "Conversatorio sobre el plan nacional de electrificación," 26.
55. Jorge Grieve, "La energía y la productividad nacional," 33.
56. For state legibility see Scott, *Seeing Like a State*.
57. Électricité de France, *Plan de electrificación nacional*, 17–18.
58. While a thorough study of the Cañón del Pato plant and the Santa Corporation is still to be written, an analysis of the project in the context of Andean climate change can be found in Carey, *In the Shadow of Melting Glaciers*. Regarding the construction of the Machu Picchu hydroelectric plant and its

role in the construction of the city of Cuzco as a tourist destination see: Rice, *Making Machu Picchu.*

59. Électricité de France, *Plan de electrificación nacional*, 68–70.
60. "Ocho mil millones de soles requeriría el plan de electrificación nacional," *El Comercio*, June 15, 1957.
61. "Electrificación del país," *Industria Peruana* 27, no. 302 (July 1957): 79–80.
62. Larsen, *Preliminary Report on Peru*, 4.
63. For a classic introductory text on the autonomy of state, see Evans, Ruschmeyer, and Skocpol, *Bringing the State Back In.*
64. Economic Commission for Latin America, *Analysis and Projections of Economic Development*, 288–91.
65. Thorp and Bertram, *Perú: 1890–1977*, 265.
66. República Peruana, *Diario de los debates de la Cámara de Diputados: Legislatura ordinaria de 1957*, vol. 4, 8.
67. República Peruana, *Diario de los debates de la Cámara de Diputados: Legislatura ordinaria de 1957*, vol. 4, 7.
68. República Peruana, *Diario de los debates de la Cámara de Diputados: Legislatura ordinaria de 1957*, vol. 4, 322–36.
69. República Peruana, *Diario de los debates de la Cámara de Diputados: Legislatura ordinaria de 1957*, vol. 4, 25–26.
70. Although commonly referred to by this name, the official title, according to the law, was the less liberal sounding Servicios Eléctricos del Estado.
71. Law No. 13979, February 5, 1962, Biblioteca del Congreso de la República, https://spij.minjus.gob.pe/Textos-pdf/Leyes/1962/Febrero/13979.pdf.
72. Manuel Prado, Speech to Congress, July 28, 1957, https://www.congreso.gob.pe/participacion/museo/congreso/mensajes/mensaje_nacion_congreso_28_julio_1957.
73. Gobierno del Perú, *Panorama de la electrificación en el Perú*, 188
74. Perú, Dirección Nacional de Estadística y Censos, *I Volumen de resultados de los censos nacionales*, 12
75. Rosenstein-Rodan, "Problems of Industrialisation of Eastern and South-Eastern Europe."
76. Rankin, "Infrastructure and the International Governance of Economic Development," 61–75. One may be more generous toward Rostow, as references to social overhead capital receive somewhat limited space in his most famous work, although he does state that "the preparation of a viable base for a modern industrial structure requires that quite revolutionary changes be brought about in two non-industrial sectors: agriculture and social overhead capital, most notably transport." Rostow, *The Stages of Economic Growth*, 25–26.
77. Larsen, *Preliminary Report on Peru*, I, 1–2.
78. Larsen, *Preliminary Report on Peru*, I.

79. Larsen, *Preliminary Report on Peru*, 36.
80. Ambassador's Impressions, Memorandum of Conversation, July 12, 1951. RG 59, 1950–54 Central Decimal File, file 823.00/7–1251, box 4591, NARA.
81. Kofas, *Foreign Debt and Underdevelopment*, 129–130.
82. Address delivered by General Manuel A. Odría before Congress, July 27, 1949, https://www.congreso.gob.pe/participacion/museo/congreso/mensajes/mensaje_nacion_congreso_27_julio_1949.
83. Office Memorandum from G Stewart Mason, December 8, 1958. Huinco Power Project, Peru, Loan 0260, P007940, box 169830B, WBGA.
84. Letter from Walter Boveri to Eugene Black, July 20, 1959. Huinco Power Project, Peru, Loan 0260, P007940, box 169830B, WBGA.
85. Office Memorandum from G. Stewart Mason, December 8, 1958. Huinco Power Project, Peru, Loan 0260, P007940, box 169830B, WBGA.
86. Letter from Walter Boveri to Eugene Black, July 20, 1959. Huinco Power Project, Peru, Loan 0260, P007940, Box 169830B, WBGA.
87. Communication from Serge Chevrier to Roger Charfounier, August 14, 1959. Huinco Power Project, Peru, Loan 0260, P007940, box 169830B, WBGA.
88. Letter from A. D. Spottswood to C. Finne, February 4, 1960. Huinco Power Project, Peru, Loan 0260, P007940, box 169830B, WBGA.
89. Report on the Salto de Huinco Hydroelectric Project, August 17, 1959. Huinco Power Project, Peru, Loan 0260, P007940, box 169830B, WBGA.
90. Ministerio de Fomento, Resolución Ministerial No. 596, June 16, 1959. Huinco Power Project, Peru, Loan 0260, P007940, box 169830B, WBGA.
91. Letter from Roger Chaufournier to Serge Chevrier, October 20, 1959. Huinco Power Project, Peru, Loan 0260, P007940, box 169830B, WBGA.
92. Letter from A. D. Spottswood to C. Finne, February 4, 1960. Huinco Power Project, Peru, Loan 0260, P007940, box 169830B, WBGA.
93. Office memorandum from C. Finne, April 27, 1960. Huinco Power Project, Peru, Loan 0260, P007940, box 169830B, WBGA.
94. Press release no. 60, June 29, 1960. Huinco Power Project, Peru, Loan 0260, P007940, box 169830B, WBGA.
95. Buse, *Huinco, 240,000 KW*, 69–70.
96. The Quechua origin of the name is somewhat confusing, roughly translating as "Puma Village Lake." The Quechua names of these bodies of water lend themselves to some interesting interpretations. Anthropologists have pointed out that the Rimac River translated as "Silent River," and pointed out that Buse's narrative changed the Rimac from "Silent River to a Working River." The current approach of anthropologists seems to indicate that these rivers should remain silent and not "work" at all. See Hommes and Boelens, "From Natural Flow to 'Working River,'" 85–95.
97. Buse, *Huinco, 240,000 KW*, 86–87.

98. "Marcapomacocha," *El Comercio*, August 16, 1957.

99. Regarding the impact on local communities, interviews carried out with residents in the area show that they had positive interactions with Pablo Boner and the company, amicable relations were maintained, and water was available for irrigation. However, the same anthropologists that carried out the interviews argue that, despite explicitly stating otherwise, the locals were mistaken, for they entered "dependent" hydraulic relations with LL&P, stating that they "still internalize and reproduce the narrative depicting the Electric Companies as saviors and light bearers." See once more: Hommes and Boelens, "From Natural Flow to 'Working River,'" 92.

100. Empresas Eléctricas Asociadas, *60 años de Empresas Eléctricas Asociadas*, 93.

101. Buse, *Huinco, 240,000 KW*, 91–92.

102. Buse, *Huinco, 240,000 KW*, 94.

103. "Arrancó Huinco-Primer Grupo," *Kilowatito*, no. 15, November 1964.

104. "El misterio biológico de la raza de bronce andina," *Kilowatito*, no. 18, October–November 1965.

105. "Collacocha en T.V.," *Kilowatito*, no. 10, December 1962.

106. Donayre Belaúnde, *Ellos también hicieron el Perú*, 99.

107. "Las obras de Huinco," *Kilowatito*, no. 9, July 1962.

108. "Fuerza para el progreso," *El Comercio*, April 3, 1965.

109. "Fuerza para el progreso," *El Comercio*, April 3, 1965.

110. Manuel Prado's ouster certainly concerned the IBRD, to the point that the military junta had to publish a document stating that the Peruvian government would continue to guarantee the loan.

111. "Inauguración Huinco," *Kilowatito*, no. 17, July 1965.

112. Inspiration for this phrase comes from: Méndez, "Incas Sí, Indios No."

113. Empresas Eléctricas Asociadas. *60 años de Empresas Eléctricas Asociadas*, 41.

114. *Electrotécnica*, no. 42, August 1965.

115. "En pocos años Lima se convirtió en la mejor iluminada de América." *La Prensa*, April 3, 1965.

116. George Young, "Power in the Andes," *International Bank Notes* 15, no. 2 (February 1961).

4. ELECTRIFICATION FROM BELOW

1. "Informe sobre el proyecto piloto de electrificación rural del Valle del Mantaro," 97.
2. Calmet, "Rural Electrification and Cooperatives," 1.
3. The best example of a discursive divide between the Mantaro Valley and the capital is Mallon, *Peasant and Nation*.
4. Burrows, "Rural Hydro-Electrification and the Colonial New Deal," 293–319; Glaser, *Electrifying the Rural American West*.

5. Engerman et al., *Staging Growth*.
6. Scott, *Seeing Like a State* and Cueto, *Excelencia científica en la periferia*.
7. See once more, Rostow, *The Stages of Economic Growth*, 4–6, and Hirschman, "Obstacles to Development," 386–88.
8. Joseph and Spenser, *In from the Cold*. For the role of experts, see Chastain and Lorek, *Itineraries of Expertise*; and Birn and Necochea López, *Peripheral Nerve*.
9. While electric cooperatives served a larger number of people in countries such as Brazil and Nicaragua, this was done through the establishment of several cooperatives, unlike in Peru, where a single cooperative was meant to serve between 100,000 and 200,000 consumers.
10. Belaúnde Terry, *Peru's Own Conquest*, 168.
11. Belaúnde Terry, *Peru's Own Conquest*, 168.
12. Castro Pozo, *Del ayllu al cooperativismo socialista*.
13. Baudin, *A Socialist Empire*, 303.
14. Belaúnde Terry, *Peru's Own Conquest*, 168.
15. Llosa Larrabure, "Cooperación Popular."
16. "*¿Que es Cooperación Popular?*"
17. Belaúnde Terry, *Peru's Own Conquest*, 167.
18. Law no. 15260, December 14, 1964, Biblioteca del Congreso de la República, https://www.leyes.congreso.gob.pe/Documentos/Leyes/15260.pdf.
19. Sharp, *U.S. Foreign Policy and Peru*, 383.
20. USOM Report on Cooperatives, July 17, 1961. RG 286, Central Subject Files 1960–1962, USAID Mission to Peru/Executive Office, box 2, NARA.
21. Law no. 9714, January 8, 1943, Biblioteca del Congreso de la República, https://www.leyes.congreso.gob.pe/Documentos/Leyes/09714.pdf.
22. Haya de la Torre, *El antiimperialismo y el APRA*, 206.
23. USOM Report on Cooperatives, July 17, 1961. RG 286, Central Subject Files 1960–1962, USAID Mission to Peru/Executive Office, box 2, NARA.
24. Mallon, *Peasant and Nation*.
25. "El desarrollo comunal del centro," *La Voz de Huancayo*, January 19, 1960. For his analysis of the impact of Western culture on the Andes, see Lizárraga, *El Perú y la cultura occidental*. Regarding the term "transculturation" see Ortiz, *Cuban Counterpoint*.
26. Long and Roberts, *Miners, Peasants and Entrepreneurs*, 27–28.
27. Long and Roberts, *Peasant Cooperation and Capitalist Expansion in Central Peru*, 4.
28. "El desarrollo comunal del centro," *La Voz de Huancayo*, January 19, 1960.
29. "La Cooperativa Comunal y la transformación socioeconómica del Perú," *La Voz de Huancayo*, June 25, 1963.
30. "La Cooperativa Comunal y la transformación socioeconómica del Perú," *La Voz de Huancayo*, June 25, 1963.

31. “Cooperativas y el Valle del Mantaro,” *La Voz de Huancayo*, July 17, 1962.
32. The Universidad Nacional del Centro—originally named the Universidad Comunal del Perú—was founded in 1959. Notable apristas, including Véliz Lizárraga, as well as Ramiro Prialé and noted geographer Javier Pulgar Vidal, played a key role in its creation.
33. Mariátegui, *Siete ensayos de interpretación de la realidad peruana*, 69–70.
34. On Muquiyauyo, see Adams, *A Community in the Andes*; Álvarez Ramos, “Muquiyauyo en el S. XX,” 263–302; Grondin, “Peasant Cooperation and Dependency,” 99–128; Plasencia Soto, “El cambio socioeconómico en las comunidades de la sierra central,” 127–40.
35. “Aumento de tarifas eléctricas,” *La Voz de Huancayo*, August 5, 1961.
36. “Centro Muquiyauyo recomienda cortar suministro eléctrico por negativa de pagos en Jauja,” *La Voz de Huancayo*, May 19, 1961.
37. “Alumbrado eléctrico de Concepción adolece de continuas interrupciones,” *La Voz de Huancayo*, Feb. 1, 1960.
38. “Problema de alumbrado en Jauja,” *La Voz de Huancayo*, May 16, 1961.
39. Électricité de France, *Plan de electrificación nacional*, 90–97.
40. Schmidt, “Political Variables and Governmental Decentralization,” 206.
41. “Obras hidroeléctricas,” *La Voz de Huancayo*, May 30, 1961.
42. Quiroz, *Corrupt Circles*, 310–11.
43. Not to be confused with the city of the same name in the Puno department.
44. “Hidroeléctrica de Pucará se entregará la próxima semana,” *La Voz de Huancayo*, March 8, 1961.
45. “Ayer se inauguró solemnemente central hidroeléctrica de Pucará,” *La Voz de Huancayo*, April 17, 1961.
46. “Hidroeléctrica de Pucará,” *La Voz de Huancayo*, April 16, 1961.
47. República Peruana, *Diario de los debates de la Cámara de Diputados: Legislatura ordinaria de 1964*, vol. 6, 275–76.
48. See: Chastain and Lorek, *Itineraries of Expertise*.
49. United States Congress, *Implementation of the Humphrey Amendment to the Foreign Assistance Act of 1961*, iii.
50. Humphrey, “U.S. Policy in Latin America,” 589.
51. Ellis, *A Giant Step*, 201–14.
52. For an early example of this dynamic see: Burrows, “Rural Hydro-Electrification and the Colonial New Deal,” 293–319.
53. Also known as the Achamayo River.
54. Gonzáles Castillo and Mendoza Rubio, “Estado y plan de desarrollo del sistema eléctrico de la Cooperativa Eléctrica Comunal del Centro Ltda. No 127,” 4–5.
55. SCIPA, *Desarrollo agrícola y económico de la zona del Mantaro en el Perú*, 17–20.
56. Melgar Lazo, *Trazos para la historia de la cooperativa eléctrica*, 8.

57. "Cooperativa de electrificación está en marcha," *La Voz de Huancayo*, November 23, 1964. For the bureaucratic and legal makeup of the cooperative, see: Eléctrica Comunal del Centro Ltda. No. 127, *Estatuto*.
58. Gonzáles Castillo and Mendoza Rubio, "Estado y plan de desarrollo del sistema eléctrico de la Cooperativa Eléctrica Comunal del Centro Ltda. No 127," 72–73.
59. "Electrificación rural," *La Voz de Huancayo*, November 23, 1964.
60. "Cooperativa de distribución de energía eléctrica," *El Comercio*, November 21, 1964.
61. "Nuevos éxitos de Cooperación Popular," *El Comercio*, November 22, 1964.
62. "Power Cooperative is Formed in Peru," *New York Times*, December 13, 1964.
63. "Pueblos del Mantaro forman cooperativa eléctrica que dará energía a 200 mil habitantes," *El Comercio*, November 20, 1964.
64. Taras Prytula (former Peace Corps volunteer) in conversation with the author, January 2022.
65. United States Congress, *Foreign Assistance Act of 1964, Hearings Before the Committee on Foreign Affairs, House of Representatives*, Part 1, 1104.
66. Taras Prytula, in conversation with the author, January 2022. Most of the NRECA reels were lost in fires or destroyed by mold over the years. However, *By the People for the People* has been digitized and is available online: https://www.youtube.com/watch?v=-V_en4eil3w&t=2s.
67. Tomás Gonzáles (former manager of the cooperative) in an email to the author, January 2022.
68. Troy Mitchell to Robert Culbertson, May 17, 1965, RG 286, Subject Files 1962–1973, USAID Mission to Peru, Private Enterprise Division, box 5, NARA.
69. Taras Prytula, in conversation with the author, January 2022
70. "Sicaya se convirtió en la ciudad mejor iluminada del valle," *La Voz de Huancayo*, July 1, 1970. Evensen was member 0914 and proudly kept his membership stub long after he returned to the United States.
71. William Evensen (former Peace Corps volunteer) in conversation with the author, December 2018.
72. Gonzáles Castillo and Mendoza Rubio, "Estado y plan de desarrollo del sistema eléctrico de la Cooperativa Eléctrica Comunal del Centro Ltda. No 127," 72–73.
73. William Evensen, in conversation with the author, December 2018.
74. For a social history of Sicaya and the role of the Baquerizo family, see Escobar, *Sicaya*.
75. Evensen, *Chug-a-luggin' & Other Acts of Andean Beer-Drinking Etiquette*.
76. Francis Dimond to Troy Mitchell, June 22, 1965. RG 286, Subject Files 1962–1973, USAID Mission to Peru, Private Enterprise Division, box 5, NARA.
77. Taras Prytula, in conversation with the author, January 2022.

78. Melgar Lazo, *Trazos para la historia de la cooperativa eléctrica*, 9; Taras Prytula, in conversation with the author, January 2022.
79. "Republica Roja de Pucutá," *La Voz de Huancayo*, June 9, 1965. For a history of the Revolutionary Left Movement, see Rubio, "Las guerrillas peruanas de 1965," 123–67; and Lust, *La lucha revolucionaria*. For the impact of the guerrillas on bilateral relations, see Walter, *Peru and the United States*.
80. Sheffield, "Peru and the Peace Corps," 359. For the viewpoint of the PCs on U.S. policy toward Peru during these years, see: Evensen, *JFK & RFK Made Me Do It*.
81. Sheffield, "Peru and the Peace Corps," 359–60.
82. "Energía eléctrica para el Valle," *La Voz de Huancayo*, December 12, 1964.
83. "Cooperativa reclama derecho de invertir 43 millones de soles," *La Voz de Huancayo*, January 16, 1967.
84. "Peruvian Visitors," *The Camden News* (Camden AK), October 3, 1970.
85. Troy Mitchell to Jim Cobb, July 12, 1967, RG 286, Subject Files 1962–1973, USAID Mission to Peru, Private Enterprise Division, box 5, NARA.
86. Troy Mitchell to Jim Cobb, July 12, 1967. RG 286, Subject Files 1962–1973, USAID Mission to Peru, Private Enterprise Division, box 5, NARA.
87. Gonzáles Castillo and Mendoza Rubio, "Estado y plan de desarrollo del sistema eléctrico de la Cooperativa Eléctrica Comunal del Centro Ltda. No 127," 347.
88. E. J. Ballard to Leon Evans, June 23, 1965. RG 286, Subject Files 1962–1973, USAID Mission to Peru, Private Enterprise Division, box 5, NARA.
89. Glaser, *Electrifying the Rural American West*.
90. Power use and member relations. Survey Trip to Peru, South America and Nicaragua, Central America. July 23–August 16, 1968. RG 286, Correspondence Relating to Cooperative Contractors, 1967–1971, Office for Private Overseas Programs/Technical Programs, box no. 4, NARA.
91. Gonzáles Castillo and Mendoza Rubio, "Estado y plan de desarrollo del sistema eléctrico de la Cooperativa Eléctrica Comunal del Centro Ltda. No 127," 185.
92. Gonzáles Castillo and Mendoza Rubio, "Estado y plan de desarrollo del sistema eléctrico de la Cooperativa Eléctrica Comunal del Centro Ltda. No 127," 187–89
93. Regarding relations with the United States, in a conversation with Ambassador Jones, Belaúnde expressed his complaints. "Belaunde said he had only eighteen months left in office. In general, he was pessimistic. He had not had the support of the United States that would have allowed him to take Peru where it had to be." Telegram from the Embassy in Peru to the State Department, February 27, 1968. RG 59, Central Files 1967–69, DEF 1 Peru, NARA, https://history.state.gov/historicaldocuments/frus1964-68v31/d502.

94. “Cooperativa Eléctrica Comunal del Centro: pasos al progreso,” *La Voz de Huancayo*, June 1, 1969.
95. Tomás Gonzáles in an email to the author, January 2022.
96. “Light of the Andes,” 1969, RG 306, Local ID 306.3324, NARA.
97. National Rural Electric Cooperative Association, *Helping Others Help Themselves*.
98. Refers to the celebration of events, sometimes historical and sometimes ahistorical, through music and dance, reflecting the particular traditions of each village.
99. Tomás Gonzáles in an email to the author, January 2022
100. “Hermanos cooperativistas,” *La Voz de Huancayo*, June 21, 1970.
101. “Luchamos por el bienestar del pueblo,” *La Voz de Huancayo*, June 25, 1970.
102. Tomás Gonzáles in an email to the author, January 2022. The story is confirmed by Taras Prytula, in conversation with the author, January 2022.
103. “A nuestros asociados y a la opinión pública,” *La Voz de Huancayo*, January 15, 1972.
104. “Cinco horas duró Asamblea de Cooperativa Eléctrica,” *La Voz de Huancayo*, January 31, 1972.
105. “Cinco horas duró Asamblea de Cooperativa Eléctrica,” *La Voz de Huancayo*, January 31, 1972.
106. Tomás Gonzáles in an email to the author, January 2022.
107. Law no. 19521, September 5, 1972, Biblioteca del Congreso de la República, https://www.leyes.congreso.gob.pe/Documentos/Leyes/19522.pdf.
108. “Del cooperativismo a la propiedad social,” *La Voz de Huancayo*, February 16, 1976.
109. “El Embajador Ruso fue declarado profesor honorario de la UNCP,” *La Voz de Huancayo*, March 22, 1976.
110. Taras Prytula in conversation with the author, January 2022.
111. Luis Carlos Arroyo (engineer for the cooperative) in conversation with the author. Lima, February 12, 2019. This perception was shared by Tomás Gonzáles.
112. “En abril la concesión del Valle del Mantaro pasará a Electroperú,” *La Voz de Huancayo*, January 4, 1976.
113. “Planta de procesamiento de postes quedará en manos de cooperativa,” *La Voz de Huancayo*, May 23, 1976.
114. “Law No. 24118, May 8, 1985,” *El Peruano*, May 9, 1985.

5. ELECTRIC REVOLUTIONS

1. “La hidroeléctrica del Mantaro,” *La Voz de Huancayo*, August 6, 1964.
2. Latham, *The Right Kind of Revolution*.
3. For studies on the Global South see: Engerman et al., *Staging Growth*.

4. McCook, *States of Nature*. Cueto, *Excelencia científica en la periferia* and *Saberes andinos*.
5. For this divide in the United States see: Needham, *Power Lines*.
6. Aguirre and Drinot, *The Peculiar Revolution*.
7. Perú. Oficina Nacional de Estadística y Censos, *Censos nacionales, VII de población, II de vivienda, 1972, Vol. 15: Departamento de Lima*, 8.
8. Ráez y Gómez, "El Mantaro y sus afluentes," 162.
9. Ráez y Gómez, "El Mantaro y sus afluentes," 173.
10. For the definition of a "creole scientist" see, McCook, *States of Nature*, 5. For the relationship among local knowledge, Western science, and the resulting scientific excellence from the periphery, see once more, Cueto, *Excelencia científica en la periferia* and *Saberes andinos*.
11. Antúnez de Mayolo, *Santiago Antúnez de Mayolo*, 10.
12. Ramírez Alzamora Cobos, *Santiago Antúnez de Mayolo*, 50.
13. Antúnez de Mayolo, *Relato de una idea*, 9.
14. Antúnez de Mayolo, *Relato de una idea*, 10.
15. Antúnez de Mayolo, *Relato de una idea*, 26.
16. Antúnez de Mayolo, *Santiago Antúnez de Mayolo*, 30–35.
17. Law No. 9089, April 27, 1940, Biblioteca del Congreso de la República, https://spij.minjus.gob.pe/Textos-PDF/Leyes/1940/Junio/09089.pdf.
18. Universidad Nacional de Ingeniería, *Foro sobre el proyecto hidroeléctrico del río Mantaro, del 27 al 31 de octubre de 1964*, 109.
19. Antúnez de Mayolo, *Relato de una idea*, 72.
20. Antúnez de Mayolo, *Relato de una idea*, 72.
21. Antúnez de Mayolo, *Memoria*, 5.
22. Yauri Montero, "Deidades panandinas del Perú antiguo en el Callejón de Huaylas," 60.
23. Antúnez de Mayolo, *Relato de una idea*, 93.
24. Antúnez de Mayolo, *La Gran Lima y la desviación del Rio Mantaro al Rimac*, 10.
25. Électricité de France, *Plan de electrificación nacional*, 90.
26. Électricité de France, *Plan de electrificación nacional*, 90.
27. "Estudio del Proyecto central hidroeléctrica de Pongor," *La Voz de Huancayo*, July 31, 1960.
28. "Japoneses retornaron ayer de Pongor," *La Voz de Huancayo*, September 18, 1960.
29. Little, *A Program for the Industrial and Regional Development of Peru*, xv.
30. Little, *A Program for the Industrial and Regional Development of Peru*, xix–xx.
31. Little, *A Program for the Industrial and Regional Development of Peru*, 61.
32.. Little, *A Program for the Industrial and Regional Development of Peru*, 64.

33. Arthur D. Little Mission Completes Economic Development Field Studies. Lima, August 18, 1960. RG 59, Central Decimal Files 1960–1963, file 823.00/8-1860, box 2381, NARA.
34. "Plan Perú-Via," *La Voz de Huancayo*, February 14, 1960.
35. Rostow, *The Stages of Economic Growth.*
36. "Comunismo y Desarrollo," *La Voz de Huancayo*, September 9, 1960.
37. "Dia de la Industria," *La Voz de Huancayo*, November 30, 1960.
38. República Peruana, *Diario de los debates de la Cámara de Diputados: Legislatura ordinaria de 1959*, vol. 1, 173.
39. República Peruana, *Diario de los debates de la Cámara de Diputados: Segunda Legislatura extraordinaria de 1960*, vol. 3, 288. For early attempts at land reform see: Pease García, "La reforma agraria peruana en la crisis del estado oligárquico," 52–53.
40. Rostow, *The Stages of Economic Growth*, 21–24.
41. República Peruana, *Diario de los debates de la Cámara de Diputados: Segunda Legislatura extraordinaria de 1960*, vol. 3, 305–6.
42. República Peruana, *Diario de los debates de la Cámara de Diputados: Legislatura ordinaria de 1961*, vol. 3, 528.
43. República Peruana, *Diario de los debates de la Cámara de Diputados: Legislatura ordinaria de 1961*, vol. 3, 534.
44. República Peruana, *Diario de los debates de la Cámara de Diputados: Legislatura ordinaria de 1960*, vol. 2, 361.
45. República Peruana, *Diario de los debates de la Cámara de Diputados: Legislatura ordinaria de 1961*, vol. 4, 43.
46. Universidad Nacional de Ingeniería, *Foro sobre el proyecto hidroeléctrico del río Mantaro*, ii.
47. República Peruana, *Diario de los debates de la Cámara de Diputados: Legislatura ordinaria de 1961*, vol. 4, 171.
48. The obsession with planning was not welcomed by everyone. David Lilienthal, the former president of the TVA, would complain years later, as he was advising on the development of hydroelectric projects in the north of the country, about "so unplanned a country as the U.S. insisting through its AID program (and the World Bank in part) on a 'national plan' before providing funds to such a country as Peru!" Lilienthal, *The Journals of David E. Lilienthal*, 74.
49. Grieve, *El proyecto Mantaro*, 5.
50. St. John, *The Foreign Policy of Peru*, 188.
51. "Lo que usted debe saber sobre la Corporación del Mantaro," *El Comercio*, March 18, 1962.
52. "La Central Hidroeléctrica del Mantaro," *La Prensa*, June 1, 1962.

53. For the notion of revolution according to Modernization Theory, see Latham, *The Right Kind of Revolution*.
54. Gobierno del Perú, *Panorama de la electrificación en el Perú*, 50–51.
55. Gobierno del Perú, *Panorama de la electrificación en el Perú*, 197.
56. Gobierno del Perú, *Plan nacional de desarrollo económico y social del Perú 1962–1971*, 233.
57. CORMAN, *Anexos al Informe de la Corporación sobre el aprovechamiento hidroeléctrico del Mantaro inferior*, annex no. 8.
58. "Huancayo debe intervenir en Corporación del Mantaro," *La Voz de Huancayo*, May 15, 1963.
59. "La central eléctrica del Mantaro y la catarata del Pongor," *La Voz de Huancayo*, May 23, 1963.
60. "La hidroeléctrica del Mantaro," *La Voz de Huancayo*, July 5, 1963.
61. Telegram from the Embassy in Lima to the Department of State, January 28. 1963. RG 59, 1960–1963 Central Decimal Files, file 823.2614/1–2863, box 2385, NARA.
62. Telegram from the Embassy in Lima to the Department of State, January 24, 1963. RG 59, 1960–1963 Central Decimal Files, file 823.26414/1–2463, box 2385, NARA.
63. Belaúnde Terry, *Peru's Own Conquest*, 47.
64. Belaúnde's decision was also influenced by a letter from President Johnson, who lauded his Amazonian highway as "inspiration for the whole hemisphere." "Mensaje de Johnson trajo a Belaúnde economista Rostow," *El Comercio*, February 6, 1966; Telegram from the Embassy in Lima to the Department of State, February 14, 1966. RG 59, Central Foreign Policy Files 1964–66, file 1-R-AID(US)-9-PERU, box 578.
65. "Surge conspiración contra hidroeléctrica del Mantaro," *La Voz de Huancayo*, November 3, 1963.
66. "Hay una bolsa de 60 millones contra Pongor," *La Voz de Huancayo*, November 4, 1963.
67. "Comunistas hostilizan las zonas que beneficiaría proyecto," *La Voz de Huancayo*, July 4, 1965.
68. "Movimiento Cívico de Defensa del Proyecto Mantaro," *La Voz de Huancayo*, January 28, 1966.
69. "£35m. contract lost to Britain," *The Times of London*, January. 31, 1966.
70. "Italia se lleva el Mantaro," *Expreso*, February 19, 1966.
71. "Italia se lleva el Mantaro," *Expreso*, February 22, 1966.
72. "Pongor cobro su víctima," *La Voz de Huancayo*, August 6, 1964.
73. "En accidente seis mueren: Pampas," *La Voz de Huancayo*, August 18, 1965.
74. Letter from Isabel Montañez to Public Notary Pedro Durand, January 3, 1967. Sección Notarial, Ex notarios de la provincia de Pampas—Tayacaja, Pedro Pablo Durand 1959–1969, book 3—file 8667952, ARJ.

75. "Huancavelica aprueba denuncia contra contratista del Mantaro," *La Voz de Huancayo*, November 17, 1963.
76. "Tayacaja busca anexo al departamento de Junín," *La Voz de Huancayo*, November 13, 1965.
77. Impresit Girola Lodigiani, IMPREGILO, *Planta Hidroeléctrica del Mantaro.*
78. Segundo Reymundo (official photographer of ELECTROPERU) in conversation with the author, December 2020.
79. Palavicino Fernández, "Labor de la educadora," 31.
80. Palavicino Fernández, "Labor de la educadora," 37.
81. "Explosivos roban a la CORMAN," *La Voz de Huancayo*, November 1, 1967.
82. "Ingeniero de GIO se ahogó en túnel de hidroeléctrica," *La Voz de Huancayo*, June 13, 1967.
83. "Cadáveres de ingenieros italianos serán trasladados a su país," *La Voz de Huancayo*, April 19, 1968; "GIE Impregilo levantó casa parroquial." *La Voz de Huancayo, September 12, 1968.*
84. Centro de Documentación Económico-Social, *Las empresas estatales*, 69.
85. Centro de Documentación Económico-Social, *Las empresas estatales*, 78.
86. "Opposition to Hydroelectric Plant Handover," *Oiga*, June 21, 1968.
87. For classic interpretations of the GRFA, see Lowenthal and McClintock, *The Peruvian Experiment Reconsidered.* For new approaches see Aguirre and Drinot, *The Peculiar Revolution.*
88. Ministerio de Energía y Minas, *Política de electricidad de la revolución peruana*, 45.
89. Ministerio de Energía y Minas, *Exposición del Ministro de Energía y Minas.*
90. For the internal dynamic of the armed forces during this period, the key text continues to be Pease García, *El ocaso del poder oligárquico.*
91. Wolfenson, *El gran desafío*, 115.
92. Corporación de Energía Eléctrica del Mantaro, *Memoria anual 1972*, 2.
93. Hidalgo, *Hidroeléctrica del Mantaro*, 119.
94. Berrios and Blasier, "Peru and the Soviet Union," 366.
95. Telegram from the Embassy in Lima to the Department of State, February 12, 1973. RG 59, Subject Numeric Files 1970–1973, file E-5-PERU, box 774, NARA.
96. North, "Orientaciones ideológicas de los dirigentes militares peruanos," 275.
97. Gobierno del Perú, *Plan nacional de desarrollo para 1971–1975*, vol. 5, 99.
98. Gobierno del Perú, *Plan nacional de desarrollo para 1971–1975*, vol. 5, 100.
99. Asociación Electrotécnica Peruana, *Anales del fórum de la ley de industria eléctrica No. 12378*, 2–3.
100. Ministerio de Energía y Minas, *Política de electricidad de la revolución peruana*, 5.
101. Ministerio de Energía y Minas, *Siete años de revolución en el sector energía y minas*, 56.

102. Telegram from Embassy in Lima to the Secretary of State, July 5, 1970. RG 59, Subject Numeric Files 1970–1973, file FSE-12-PERU, box 997, NARA.
103. "Peruanizan la Electricidad," *La Voz de Huancayo*, September 14, 1972.
104. "130 funcionarios Italos verán inauguración del Mantaro," *La Prensa*, October 4, 1973.
105. "Central del Mantaro comenzó a funcionar con primera turbina," *La Prensa*, October 7, 1973.
106. "Central del Mantaro, triunfo de la Revolución Peruana: F. Maldonado," *Expreso*, October 7, 1973.
107. "Central del Mantaro, triunfo de la Revolución Peruana: F. Maldonado," *Expreso*, October 7, 1973.
108. "Obra es fruto señero de un gran pueblo," *Expreso*, October 7, 1973.
109. Gobierno Revolucionario de la Fuerzas Armadas, *Revolución con electrificación*, 3.
110. UNICEF, *Servicios básicos*, 339.
111. Collier, *Squatters and Oligarchs*, 98. For Velasco's policy toward the barriadas, see also: Dietz, *Poverty and Problem-Solving under Military Rule.*
112. UNICEF, *Servicios básicos*, 340. The Junta Nacional de Vivienda was established in 1962 and was the successor of the Comisión Nacional de Vivienda of 1945. Both had the aim of improving housing conditions in Lima.
113. Empresas Eléctricas Asociadas, *60 años de Empresas Eléctricas Asociadas*, 100.
114. Dietz, *Poverty and Problem-Solving Under Military Rule*, 232.
115. Oficina Nacional de Desarrollo de los Pueblos Jóvenes, *Anuario 1969–1970*, 32.
116. Letter from Mario Julca Gallardo, September 24, 1973. Colección Energía y Minas, folder 3132 (85)-319544, box 23, AGN.
117. Letter from Tayacaja Coop, November 14, 1973. Colección Energía y Minas, folder 3132 (100)-320661, box 24, AGN.
118. Letter from the Comité Vecinal de Villa el Salvador, Gr. 8, Lote 1, July 5, 1972. Sección Energía y Minas, folder 3132 (13)–310994. box 20, AGN.
119. Hidalgo, *Hidroeléctrica del Mantaro*, 130.
120. Kuczynski Godard, *Memoria de Energía y Minas 1980–1982*, 68.

EPILOGUE

1. Ministerio de Energía y Minas, *Energía eléctrica y desarrollo*, 18
2. "El porqué Lima se va muriendo de sed," *La Razón*, April 13, 2004. Hydroelectric projects were resumed in earnest in the twenty-first century. In 2016, the Central Hidroeléctrica Cerro del Águila, the country's second largest, which is also fed by the waters of the Mantaro, was inaugurated. However, few Peruvians paid attention, as hydroelectric development no longer captured the nation's imagination.

3. Pease García, *Un perfil del proceso político peruano.*
4. Kuczynski Godard, *Memoria de Energía y Minas*, vol. 1, 63.
5. Kuczynski Godard, *Memoria de Energía y Minas*, vol. 1, 18–19.
6. Kent, "Geographical Dimensions of the Shining Path Insurgency in Peru," 446.
7. Comisión de la Verdad y Reconciliación, *Hatun Willakuy*, vol. 4, 180.
8. DESCO, *Violencia política en el Perú, 1980–1988*, vol. 1, 169
9. Kent, "Geographical Dimensions of the Shining Path Insurgency in Peru," 446.
10. Comisión de la Verdad y Reconciliación, *Hatun Willakuy*, vol. 4, 219.
11. Comisión de la Verdad y Reconciliación, *Hatun Willakuy*, vol. 4, 219.
12. Bonfiglio, *Historia de la electricidad en Lima*, 90.
13. "Terroristas dinamitaron 6 torres de energía eléctrica," *El Comercio*, July 22, 1983.
14. "Voladura de torres de alta tensión paralizó producción, afectó a radios, grifos y diarios," *El Comercio*, July 23, 1983.
15. "Voladura de torres de alta tensión paralizó producción, afectó a radios, grifos y diarios," *El Comercio*, July 23, 1983.
16. "Dos PIP malheridos y 10 torres voladas," *El Observador*, October 16, 1983.
17. DESCO, *Violencia política en el Perú, 1980–1988*, vol. 1, 47.
18. Comisión de la Verdad y Reconciliación. *Hatun Willakuy*, vol. 4, 215.
19. Comisión de la Verdad y Reconciliación. *Hatun Willakuy*, vol. 4, 424.
20. Tarazona-Sevillano, "The Organization of Shining Path," 196.
21. DESCO, *Violencia política en el Perú, 1980–1988*, vol. 1, 122.
22. It is even more ironic that in 1986 Sendero inmates in Lima's most well known prison, El Frontón, presented a list of demands that included "electric power supply 24 hours per day" as well as the "reparation of the water, sewer and electric system in Callao and Lurigancho." Communist Party of Peru, "Demands of the Prisoners of War of the Shining Trenches of Combat of Fronton, Lurigancho and Callao."
23. Hidalgo, *Tejedores de luz*, 88.
24. Jaime Guerra (former director of SEIN) in conversation with the author, January 2019.
25. Hidalgo, *Tejedores de luz*, 83. For a new approximation to the El Niño phenomenon see: Puente, "Making Peru's Sendero Luminoso."
26. Ministerio de Energía y Minas, "Central hidroeléctrica de Callahuanca reinicia operaciones y contribuye a fortalecer el sistema eléctrico del país," November 22, 2019, https://www.gob.pe/institucion/minem/noticias/69382-central-hidroelectrica-de-callahuanca-reinicia-operaciones-y-contribuye-a-fortalecer-el-sistema-electrico-del-pais.
27. United Nations Framework Convention on Climate Change, "How Hydropower Can Help Climate Action," November 21, 2018, https://unfccc.int/news/how-hydropower-can-help-climate-action.

28. Julia Cuadros, "China y el monopolio de la electricidad en Lima," February 14, 2024, https://cooperaccion.org.pe/opinion/china-y-el-monopolio-de-la-electricidad-en-lima/.
29. "Huancavelica: 400 manifestantes amenazan con tomar la central hidroeléctrica más importante del país," RPP, December 14, 2022, https://rpp.pe/peru/huancavelica/huancavelica-400-manifestantes-amenazan-con-tomar-la-central-hidroelectrica-mas-importante-del-pais-noticia-1453600.

BIBLIOGRAPHY

ARCHIVES AND COLLECTIONS

College Park, Maryland, USA

National Archives and Records Administration

Record Group 38 Office of Naval Intelligence (RG 38)

Record Group 59 Department of State (RG 59)

Record Group 286 Agency for International Development (RG 286)

Record Group 306 United States Information Agency (RG 306)

Washington DC, USA

Library of Congress

The World Bank Group Archives (WBGA)

New York, USA

New York Public Library

Science, Industry and Business Library

Austin, Texas, USA

University of Texas Libraries

Huancayo, Peru

Archivo Regional de Junín (ARJ)

Biblioteca Municipal Alejandro Deústua

Lima, Peru

Archivo del Ministerio de Transportes y Telecomunicaciones

Colección Sub-Dirección de Electricidad (MTC-SDE)

Archivo del Museo de la Electricidad

Archivo General de la Nación

Colección Ministerio de Energía y Minas (AGN-EyM)

Asociación Electrotécnica Peruana

Biblioteca Nacional del Perú

Biblioteca del Congreso del Perú

Pontificia Universidad Católica del Perú Libraries

Instituto Riva Agüero

Sociedad Geográfica de Lima

INTERVIEWS

Luis Carlos Arroyo—engineer, Cooperativa Eléctrica Comunal del Centro

William Evensen—Peace Corps volunteer

Tomás Gonzáles—manager, Cooperativa Eléctrica Comunal del Centro
Jaime Guerra—former director, Comité de Operación Económica del Sistema Interconectado
Taras Prytula—Peace Corps volunteer
Segundo Reymundo—Official photographer, ELECTROPERU
Anibal Esteban Tomecich—engineer, Cerro de Pasco Corporation
Azi Wolfenson—former head of ELECTROPERU

PUBLISHED WORKS

Adams, Jorge. "Caudal, procedencia y distribución de aguas de los departamentos de Lima e Ica." *Boletín del Cuerpo de Ingenieros de Minas del Perú*, no. 37 (1906).

Adams, Richard N. *A Community in the Andes: Problems and Progress in Muquiyauyo*. Seattle: University of Washington Press, 1959.

Aguirre, Carlos, and Paulo Drinot, eds. *The Peculiar Revolution: Rethinking the Peruvian Experiment under Military Rule*. Austin: University of Texas Press, 2017.

Alcalde, Javier Gonzalo. *The Idea of Third World Development: Emerging Perspectives in the United States and Britain: 1900–1950*. Lanham NY: University Press of America, 1987.

Alfonso Pezet, Federico. "Estudio de la colonización del Perú, bajo el punto de vista práctico." *Boletín de la Sociedad Geográfica de Lima*, vol. 4 (1895): 121–32.

Álvarez Ramos, José Luis. "Muquiyauyo en el S. XX: exploraciones en su historia y las relaciones entre el Estado y la Comunidad desde un estudio de la primera empresa eléctrica comunal." In *Pueblos del Hatun Mayu: historia, Arqueología y Antropología en el valle del Mantaro*, edited by José Luis Álvarez Ramos, Carlos H. Hurtado Ames, and Manuel F. Perales Munguía, 263–302. Lima: CONCYTEC, 2011.

Anand, Nikhil, Akhil Gupta, and Hannah Appel, eds. *The Promise of Infrastructure*. Durham NC: Duke University Press, 2018.

Antúnez de Mayolo, Santiago. *La génesis de los servicios eléctricos de Lima*. Lima: Impr. E.Z. Casanova, 1929.

———. *La Gran Lima y la desviación del Rio Mantaro al Rimac*. Lima: Sociedad de Ingenieros del Perú, 1953.

———. *Memoria: trazo preliminar de la Central Eléctrica del Pongor en el Mantaro*. Lima: Rímac, 1945.

———. *Relato de una idea a su realización, o, la Central Hidroeléctrica del Cañón del Pato*. Lima: Editorial Médica Peruana, 1957.

———. *Santiago Antúnez de Mayolo*. Lima: Juan Mejía Baca, 1966.

———. *Santiago Antúnez de Mayolo: vida y obra*. Edited by Santiago Erick Antúnez de Mayolo. Lima: Talleres IBEGRAF, 2006.

Arana, Víctor M. *La electricidad en el hogar: manual explicativo sobre utensilios y máquinas eléctricas para uso doméstico*. Lima: n.p., 1929.

Asociación Electrotécnica Peruana. *Anales del fórum de la Ley de industria eléctrica No. 12378*. Lima: AEP, 1969.

———. *Libro de oro: 1943–1993*. Lima: SAGSA, 1993.

Bardella, Gianfranco. *Un siglo en la vida económica del Perú 1889–1989*. Lima: Banco de Crédito del Perú, 1989.

Barrera, G. "Líneas de transmisión a 64 KV Moyopampa-Lima," *Electrotécnica*, no. 7 (January 1954): 9–17.

Basadre, Jorge. *Historia de la República del Perú*. Lima: Editora El Comercio S. A., 2005.

———. *Perú: problema y posibilidad*. Lima: Banco Internacional del Perú, 1978.

Baudin, Louis. *A Socialist Empire: The Incas of Peru*. Princeton NJ: D. Van Nostrand Company, 1961.

Beals, Carlton. *The Coming Struggle for Latin America*. Philadelphia: J. B. Lippincott Company, 1938.

Belaúnde Terry, Fernando. *La conquista del Perú por los peruanos*. Lima: Minerva, 1959.

———. *Peru's Own Conquest*. Lima: American Studies Press, 1965

Berrios, Ruben, and Cole Blasier. "Peru and the Soviet Union (1969–1989): Distant Partners." *Journal of Latin American Studies* 23, no. 2 (May 1991): 365–84.

Bethell, Leslie, and Ian Roxborough, eds. *Latin America between the Second World War and the Cold War: Crisis and Containment, 1944–1948*. Cambridge UK: Cambridge University Press, 1992.

Birn, Anne-Emanuelle, and Raúl Necochea López, eds. *Peripheral Nerve: Health and Medicine in Cold War Latin America*. Durham NC, Duke University Press, 2020.

Blanc, Jacob. *Before the Flood: The Itaipu Dam and the Visibility of Rural Brazil*. Durham NC: Duke University Press, 2019.

Boner, Pablo. "El planeamiento de las obras eléctricas." *Electrotécnica*, no. 4 (April 1953): 9–13.

Boner, Pablo, and Giulio Donizetti. *Los servicios eléctricos*. Lima: Libr. Internacional del Perú, 1949.

Bonfiglio, Giovanni. *Historia de la electricidad en Lima: noventa años de modernidad*. Lima: Museo de la Electricidad, 1997.

———. *Los italianos en la sociedad peruana: una visión histórica*. Lima: Asociación Italianos del Perú, 1993.

Bonilla, Heraclio. *Guano y burguesía en el Perú*. Lima: Instituto de Estudios Peruanos, 1984.

Boveri, Walter. *Juan Carosio, 1876–1959: homenaje rendido a la memoria del Ing. J. Carosio, por el Dr. W. Boveri en la Asamblea general ordinaria de la Sociedad suizo-americano de electricidad "Samelec" y de la Sociedad sudamericana de electricidad "Sudalectra," el 29 de julio de 1959*. Lima: Tall. Gráf. Cecil, 1960.

Bueno Ortiz, Armando. "Algunos aspectos geopolíticos del Perú en la defensa nacional." *Peruanidad* 20, vol. 4 (November–December 1944): 1539–48.

Burrows, Geoff. "Rural Hydro-Electrification and the Colonial New Deal: Modernization, Experts, and Rural Life in Puerto Rico, 1935–1942." *Agricultural History* 91, no. 3 (2017): 293–319.

Buse, Hermann. *Callahuanca, 1938–1963*. Lima: EEAA, 1963.

———. *Huinco. 240,000 KW. Historia y geografía de la electricidad en Lima*. Lima: Talleres Gráficos P. L. Villanueva, 1965.

By the People for the People: The Rural Electric Story. Produced by the National Rural Electric Cooperative Association. 26 min. 1955.

Calmet, Mario. "Electrificación rural y cooperativas." In *Anales del fórum de la ley de industria eléctrica No. 12378*, Asociación Electrotécnica Peruana, 1–11. Lima: AEP, 1969.

Cámara de Comercio Suiza en el Perú. *La presencia suiza en el Perú*. Lima: Cámara de Comercio Suiza en el Perú, 1991.

Carey, Mark. *In the Shadow of Melting Glaciers: Climate Change and Andean Society*. Oxford UK: Oxford University Press, 2010.

Carranza, Luis, *Boletín de la Sociedad Geográfica de Lima*, vol. 1 (1892): 1–2.

Carrasco Valencia, Alfonso. *La electricidad en el Perú: política estatal y electrificación rural*. Lima: ITDG, 1990.

Castillo Paulino, Luis. *Seguridad y salud en el trabajo en la industria eléctrica: evolución de la industria y las luchas contra la nocividad en el Perú, 1886–1996*. Lima: Instituto de Estudios Sindicales, 2009.

Castro Pozo, Hildebrando. *Del ayllu al cooperativismo socialista*. Lima: Cía. Enrique Bustamante y Ballivian, 1936.

Cavers, David F., and James R. Nelson. *Electric Power Regulation in Latin America*. Baltimore: Johns Hopkins University Press, 1959.

Centeno, Miguel Angel. *Blood and Debt: War and the Nation-State in Latin America*. University Park: Pennsylvania State University Press, 2002.

Centro de Documentación Económico-Social. *Las empresas estatales en el Perú*. Lima: Centro de Documentación Económico-Social, 1965.

Chanduví, Luis. *El APRA por dentro: lo que hice, lo que vi, y lo que sé, 1931–1957*. Lima: Copias e Impresiones, 1988.

Chastain, Andra B., and Timothy W. Lorek, eds. *Itineraries of Expertise: Science, Technology, and the Environment in Latin America's Long Cold War*. Pittsburgh: University of Pittsburgh Press, 2020.

Chiozza Money, Leo George. "Why Should Scotland Lose Population?" *Peru Today* 4, no. 3 (June 1912): 146, 150.

Chikowero, Moses. "Subalternating Currents: Electrification and Power Politics in Bulawayo, Colonial Zimbabwe, 1894–1939." *Journal of Southern African Studies* 33, no. 2 (2007): 287–306.

Ciccarelli, Orazio A. "Fascism and Politics in Peru during the Benavides Regime, 1933–39: The Italian Perspective." *The Hispanic American Historical Review* 70, no. 3 (August 1990): 405–32.

———. "Fascist Propaganda and the Italian Community in Peru during the Benavides Regime, 1933–39." *Journal of Latin American Studies* 20, no. 2 (1988): 361–88.

———. "La crisis etíope y la liga de naciones en la prensa peruana." *Histórica* 12, no. 1 (1988): 35–48.

Clayton, Lawrence A. *Grace: W.R. Grace & Co., the Formative Years: 1850–1930*. Ottawa IL: Jameson Books, 1985.

Coleman, Leo. *A Moral Technology: Electrification as Political Ritual in New Delhi*. Ithaca NY: Cornell University Press, 2017.

Collier, David. *Squatters and Oligarchs: Authoritarian Rule and Policy Change in Peru*. Baltimore: Johns Hopkins University Press, 1976.

Comisión de la Verdad y Reconciliación. *Hatun Willakuy: versión abreviada del Informe Final de la Comisión de la Verdad y Reconciliación*. Tomo II. Lima: CVR, 2008.

———. *Hatun Willakuy: versión abreviada del Informe Final de la Comisión de la Verdad y Reconciliación*, Tomo IV. Lima: CVR, 2008.

Communist Party of Peru. *Collected Works of the Communist Party of Peru*. Vol. 1, *1968–1987*. n.p: Foreign Languages Press, 2016.

Communist Party of Peru, "Demands of the Prisoners of War of the Shining Trenches of Combat of Fronton, Lurigancho and Callao." In *Collected Works of the Communist Party of Peru*. Vol. 1, *1968–1987*. Published by Author, 2016.

Contreras, Carlos, ed. *Historia económica del Perú central: ventajas y desafíos de estar cerca de la capital*. Lima: Instituto de Estudios Peruanos, 2022.

Contreras, Carlos. "La economía del transporte en el Perú, 1800–1914." *Apuntes* 66 (2010): 59–81.

———. *Mineros y campesinos en los Andes: mercado laboral y economía campesina en la sierra central siglo XIX*. Lima: Instituto de Estudios Peruanos, 1987.

Contreras, Carlos, and Marcos Cueto. "Caminos, Ciencia y Estado en el Perú, 1850–1930." *História, Ciências, Saúde* 15, no. 3 (2008): 635–55.

Contreras, Carlos, and Paulo Drinot, "The Great Depression in Peru." In *The Great Depression in Latin America*, edited by Paulo Drinot and Alan Knight, 102–28. Durham NC: Duke University Press, 2014.

"Conversatorio sobre el plan nacional de electrificación." *Electrotécnica*, no. 17–18 (July–December 1956): 25–27.

Coopersmith, Jonathan. *The Electrification of Russia, 1880–1926*. Ithaca NY: Cornell University Press, 1992.

CORMAN, *Anexos al Informe de la Corporación sobre el aprovechamiento hidroeléctrico del Mantaro inferior*, Lima: CORMAN, 1964, annex no. 8.

Coronel Zegarra, Enrique. "El ferrocarril de Paita al Marañón," *Boletín de la Sociedad Geográfica de Lima*, vol. 17 (1905): 447–68.

Corporación de Energía Eléctrica del Mantaro. *anexos al Informe de la Corporación sobre el aprovechamiento hidroeléctrico del Mantaro inferior*. Lima: CORMAN, 1964.

———. *Memoria anual 1963*. Lima: CORMAN, 1963.

———. *Memoria anual 1964*. Lima: CORMAN, 1964.

———. *Memoria anual 1966*. Lima: CORMAN, 1966.

———. *Memoria anual 1967*. Lima: CORMAN, 1967.

———. *Memoria anual 1968*. Lima: CORMAN, 1968.

———. *Memoria anual 1970*. Lima: CORMAN, 1970.

———. *Memoria anual 1972*. Lima: CORMAN, 1972.

Cronon, William. *Nature's Metropolis: Chicago and the Great West*. New York: W. W. Norton and Company, 1991.

Cuadros, Julia. "China y el monopolio de la electricidad en Lima." February 14, 2024. CooperAcción. https://cooperaccion.org.pe/opinion/china-y-el-monopolio-de-la-electricidad-en-lima/.

Cueto, Marcos. "Apogeo y crisis de la Sociedad Geográfica de Lima: 1888–1940." *Acta Hispanica ad Medicinae Scientiarumque Historiam Illustrandam* 12 (1992): 35–45.

———. *Excelencia científica en la periferia: actividades científicas e investigación biomédica en el Perú: 1890–1950*. Lima: GRADE, 1989.

———. *Saberes andinos: ciencia y tecnología en Bolivia, Ecuador y Perú*. Lima: Instituto de Estudios Peruanos, 1995.

Cushman, Gregory. *Guano and the Opening of the Pacific World: A Global Ecological History*. New York: Cambridge University Press, 2013.

Dargent, Eduardo. *Technocracy and Democracy in Latin America: The Experts Running Government*. Cambridge UK: Cambridge University Press, 2014.

Dargent Chamot, Eduardo. *El billete en el Perú*. Lima: Banco Central de Reserva del Perú, 1979.

de la Riva-Agüero y Osma, José. *Obras completas de José de la Riva-Agüero*. Vol. 13, *Epistolario Baca—Byrne*. Lima: PUCP-Instituto Riva-Agüero, 1996.

———. *Paisajes peruanos*. Lima: Imprenta Santa María, 1955.

de Vivero, León. *Avance del imperialismo fascista en el Perú*. México: Manuel Arévalo, 1938.

Delgado, Eulogio. "Memoria correspondiente al año 1905." *Boletín de la Sociedad Geográfica de Lima*, vol. 18 (1905): 3–28.

Der, Jennifer. *The Lived Nile: Environment, Disease, and Material Colonial Economy in Egypt*. Stanford: Stanford University Press, 2019.

DESCO. *Violencia política en el Perú, 1980–1988*, vol. 1. Lima: DESCO, 1989.

Dietz, Henry A. *Poverty and Problem-Solving under Military Rule: The Urban Poor in Lima, Peru*. Austin: University of Texas Press, 1980.

"Discurso del Ing. Jorge Grieve en el XV Aniversario de la A.E.P." *Electrotécnica*, no. 23–24 (January–June 1958): 26–29.

Donayre Belaúnde, Jorge. *Ellos también hicieron el Perú*. Lima: Augusto Elmore–Southern Peru, 1990.

Drake, Paul. "International Crises and Popular Movements in Latin America: Chile and Peru from the Great Depression to the Cold War." In *Latin America in the 1940s: War and Postwar Transitions*, edited by David Rock, 110–36. Berkeley: University of California Press, 1994.

———. *The Money Doctor in the Andes: The Kemmerer Missions, 1923–1933*. Durham NC: Duke University Press, 1989.

Drinot, Paulo. *The Allure of Labor: Workers, Race, and the Making of the Peruvian State*, Durham NC: Duke University Press, 2011.

———, ed. *Peru in Theory*. Oxford UK: Palgrave Macmillan, 2014.

Drinot, Paulo, and Alan Knight, eds. *The Great Depression in Latin America*. Durham NC: Duke University Press, 2014.

Durand, Francisco. *La década frustrada: los industriales y el poder, 1970–1980*. Lima: DESCO, 1982.

Economic Commission for Latin America. *Analysis and Projections of Economic Development*. Vol. 6, *The Industrial Development of Peru*. Mexico Distro Federal: United Nations Publications, 1959.

Eléctrica Comunal del Centro Ltda. No. 127, *Estatuto*. Huancayo: Editorial Sebastián. Lorente, n.d.

"El plan quinquenal para la industria del estado," *Amauta*, no. 27 (November–December 1929), 4–9.

Électricité de France. *Plan de electrificación nacional*. Paris: J&R Sennac Imprimeurs, 1957.

Ellis, Clyde T. *A Giant Step*. New York: Random House, 1966.

Empresas Eléctricas Asociadas. *60 años de Empresas Eléctricas Asociadas*. Lima: Tip. Santa Rosa, 1966.

———. *Electricidad para la Gran Lima*. Lima: Emp. Graf. T. Scheuch, 1952.

———. *Los servicios eléctricos en el Callao: la reforma y quince años de antecedentes*. Lima: Empresas Eléctricas Asociadas, 1933.

———. *Memoria del directorio 1933*. Lima: LL&P, 1934.

———. *Memoria del directorio Ejercicio 1939*. Lima: LL&P, 1940.

———. *Memoria del directorio: Ejercicio 1948*. Lima: Impr. Empresas Eléctricas, 1949.

———. *Memoria del directorio: Ejercicio 1950*. Lima: Impr. Empresas Eléctricas, 1951.

———. *Memoria del directorio: Ejercicio 1955*. Lima: Impr. Empresas Eléctricas, 1956.

———. *Memoria del directorio Ejercicio 1962*. Lima: Impr. Empresas Eléctricas, 1963.

———. *Souvenir of the Empresas Eléctricas Asociadas (Lima Light, Power and Tramways Co.)*. Lima: M. Moral, 1913.

Empresas Eléctricas Asociadas—Hidrandina. *Energía eléctrica para la Gran Lima 1955/1956*. Lima: Emp. Graf. T. Scheuch, 1955.

———. *Luz y fuerza para la Gran Lima*. Lima: Emp. Graf. T. Scheuch, 1958.

Engerman, David C., Nils Gilman, Mark H. Haefele, and Michael E. Latham, eds. *Staging Growth: Modernization, Development, and the Global Cold War*. Amherst: University of Massachusetts Press, 2003.

Enock, Charles Reginald. *The Andes and the Amazon: Life and Travel in Peru*, 4th ed. London: T. Fisher Unwin, 1910.

Escobar, Gabriel. *Sicaya: cambios culturales en una comunidad mestiza andina*. Lima: Instituto de Estudios Peruanos, 1973.

Evans, Peter. *Embedded Autonomy: States and Industrial Transformation*. Princeton NJ: Princeton University Press, 1995.

Evans, Peter, Dietrich Ruschmeyer, and Theda Skocpol, eds. *Bringing the State Back In*. Cambridge UK: Cambridge University Press, 1985.

Evensen, William. *Chug-a-luggin' & Other Acts of Andean Beer-Drinking Etiquette*. 2019 (unpublished memoir).

———. *JFK & RFK Made Me Do It: 1960–1968*. Columbia SC: Peace Corps Writers and Constitutional Capers, 2021.

Federación de Motoristas, Conductores y Anexos con las Empresas Eléctricas asociadas y la Cía. Nacional de Tranvías S.A. *Pactos colectivos suscritos entre la Federación de Motoristas, Conductores y Anexos con las Empresas Eléctricas asociadas y la Cía. Nacional de Tranvías S.A.* (Lima: Imp. El Triunfo, 1936).

Fitzgerald, E. V. K. *The Political Economy of Peru 1956–68: Economic Development and the Restructuring of Capital*. Cambridge UK: Cambridge University Press, 1979.

Flores Galindo, Alberto. *In Search of an Inca: Identity and Utopia in the Andes*. Cambridge UK: Cambridge University Press, 2010.

———. *Los mineros de la Cerro de Pasco, 1900–1930: un intento de caracterización social*. Lima: Pontificia Universidad Católica del Perú, 1974.

Folch, Christine. *Hydropolitics: The Itaipu Dam, Sovereignty, and the Engineering of Modern South America*. Princeton NJ: Princeton University Press, 2019.

Gálvez, José. *Obras Completas*, vol. 2. Lima: Okura Editores, 1985.

Gálvez, José. *Obras Completas*, vol. 4. Lima: Okura Editores, 1985.

García Calderón, Francisco. *América Latina y el Perú del novecientos: Antología de textos*. Lima: UNMSM—COFIDE, 2003.

———. *El Perú contemporáneo*. Lima: Banco Internacional del Perú, 1981.

Garland, Alejandro. "Importancia de los ríos peruanos." *Boletín de la Sociedad Geográfica de Lima*, vol. 27 (1911): 95–107.

Gerschenkron, Alexander. *Economic Backwardness in Historical Perspective*. Cambridge MA: Belknap Press of Harvard University Press, 1962.

Gilman, Nils. *Mandarins of the Future: Modernization Theory in Cold War America*. Baltimore: Johns Hopkins University Press, 2003.

Glaser, Leah S. *Electrifying the Rural American West: Stories of Power, People, and Place*. Lincoln: University of Nebraska Press, 2009.

Gobierno del Perú. *Panorama de la electrificación en el Perú: realizaciones y proyectos durante los dos gobiernos del Dr. Manuel Prado*. Lima: Gobierno del Perú, 1962.

———. *Plan nacional de desarrollo económico y social del Perú 1962–1971*. Lima: BCR, 1962.

———. *Plan nacional de desarrollo para 1971–1975*. Vol. 5, *Plan de electricidad*. Lima: Presidencia de la República, 1971.

Gobierno Revolucionario de la Fuerzas Armadas. *Revolución con electrificación*. Lima: Gobierno del Perú, 1974.

Gonzáles Castillo, Tomás Florencio, and Tomás Martín Mendoza Rubio. "Estado y plan de desarrollo del sistema eléctrico de la Cooperativa Eléctrica Comunal del Centro Ltda. No. 127." Undergraduate dissertation, Universidad Nacional de Ingeniería, 1973.

Gooday, Graeme. *Domesticating Electricity: Technology, Uncertainty, and Gender, 1880–1914*. London: Pickering & Chatto, 2008.

Gootenberg, Paul. *Andean Cocaine: The Making of a Global Drug*. Chapel Hill: University of North Carolina Press, 2008.

———. *Between Silver and Guano: Commercial Policy and the State in Postindependence Peru*. Princeton NJ: Princeton University Press, 1989.

———. "Hijos of Dr. Gerschenkron: 'Latecomer' Conceptions in Latin American Economic History." In *The Other Mirror: Grand Theory through the Lens of Latin America*, edited by M. Centeno and F. López-Alves, 57–80. Princeton NJ: Princeton University Press, 2001.

———. *Imagining Development: Economic Ideas during Peru's "Fictitious Prosperity" of Guano, 1840–1880*. Berkeley: University of California Press, 1993.

———. "Order(s) and Progress in Developmental Discourse: A Case of Nineteenth-Century Peru." *Journal of Historical Sociology* 8, no. 2 (1995): 111–35.

Grieve, Jorge. *El proyecto Mantaro: carta abierta del Ing. Jorge Grieve al Sr. Ernesto Ferreyros*. Lima: Industrialgráfica, 1965.

———. "La energía y la productividad nacional." *Electrotécnica*, no. 21–22 (July–December 1957): 33–38.

———. "Plan de electrificación nacional." *Electrotécnica*, no. 3 (January 1953): 21–26.

Grondin, Marcelo. "Peasant Cooperation and Dependency: The Case of the Electricity Enterprises of Muquiyauyo." In *Peasant Cooperation and Capitalist Expansion in Central Peru*, edited by Norman Long and Bryan Roberts, 99–128. Austin: University of Texas Press, 1979.

Guarini, Emilio. *El porvenir de la industria eléctrica en el Perú*. Lima: Imprenta de la Escuela de Ingenieros, 1907.

———. "La moderna enseñanza técnica y el porvenir de la enseñanza obrera en el Perú." *Informaciones y memorias de la Sociedad de Ingenieros del Perú*, vol. 9 (October 1907), 211–29.

Guerra, Margarita. *Manuel A. Odría*. Lima: Brasa, 1994.

Guizado Mercado, Yenisa, "De la palabra a la mano: la formación de la Educación Técnica en el Perú, 1864–1915." *ISHRA, Revista Del Instituto Seminario De Historia Rural Andina* 1, no. 9 (2022): 71–84.

Guizado Mercado, Yenisa, and José Ragas. "The National School of Electricity and Early Technical Education in Peru." *Diseña*, no. 18 (January 2021), article 15.

Hager, Willi. *Hydraulicians in Europe 1800–2000*, vol. 2. Boca Raton: CRC Press, 2014.

Harrison-Moore, Abigail, and R. W. Sandwell, eds. *In a New Light: Histories of Women and Energy*. Montreal: McGill-Queen's University Press, 2021.

Harvey, Penelope, Casper Jensen, and Atsuro Morita, eds. *Infrastructures and Social Complexity: A Companion*. London: Routledge, 2017.

Harvey, Penny, and Hannah Knox. *Roads: An Anthropology of Infrastructure and Expertise*. Ithaca NY: Cornell University Press, 2015.

Hausman, William J., Peter Herner, and Mira Wilkins. *Global Electrification: Multinational Enterprise and International Finance in the History of Light and Power, 1878-2007*. New York: Cambridge University Press, 2008.

Haworth, Nigel. "Peru." In *Latin America between the Second World War and the Cold War: Crisis and Containment, 1944–1948*, edited by Leslie Bethell and Ian Roxborough, 170–89. Cambridge UK: Cambridge University Press, 1992.

Haya de la Torre, Víctor Raúl. *El antiimperialismo y el APRA*. Lima: Fondo Editorial del Congreso, 2010.

Hiatt, Willie. *The Rarified Air of the Modern: Airplanes and Technological Modernity in the Andes*. Oxford UK: Oxford University Press, 2016.

Hidalgo, Neydo. *55 años del Cañón del Pato: una proeza en los Andes: historia de la Central Hidroeléctrica del Cañón del Pato*. Lima: Duke Energy Perú, 2013.

———. *Hidroeléctrica del Mantaro: el arte de hacer luz*. Lima: Electroperú, 2010.

———. *La profesión civilizadora: forjadores de la ingeniería eléctrica en el Perú.* Lima: Capítulo de Ingeniería Eléctrica–AEP.

———. *Tejedores de luz: homenaje a los forjadores de la transmisión eléctrica en el Perú, 1887–2007.* Lima: Red de Energía del Perú, 2007.

"High-Tension Energy Transmission in Peru." *Electrical World* 51, no. 5 (February 1908): 223–25.

Hill, Jonathan. "Circuits of State: Water, Electricity, and Power in Chihuahua, 1905–1936." *Radical History Review* 127 (2017): 13–38.

Hirschman, Albert O. "Obstacles to Development: A Classification and a Quasi-Vanishing Act." *Economic Development and Cultural Change* 13, no. 4 (1965): 385–93.

———. *The Strategy of Economic Development.* New Haven CT: Yale University Press, 1958.

Hommes, Lena, and Rutgerd Boelens. "From Natural Flow to 'Working River': Hydropower Development, Modernity and Socio-territorial Transformations in Lima's Rímac Watershed." *Journal of Historical Geography* 62 (October 2018): 85–95.

Hughes, Thomas. *Networks of Power: Electrification in Western Society, 1880–1930.* Baltimore: Johns Hopkins University Press, 1983.

Humphrey, Hubert. "U.S. Policy in Latin America." *Foreign Affairs* 42, no. 4 (July 1964): 585–601.

Impresit Girola Lodigiani, IMPREGILO. *Planta Hidroeléctrica del Mantaro.* Perú: Corporación de Energía Eléctrica del Mantaro, 1966.

"Informe sobre el proyecto piloto de electrificación rural del Valle del Mantaro." *Electrotécnica*, no. 52 (1972): 97–106.

Innis, Harold. *The Fur Trade in Canada: An Introduction to Canadian Economic History.* New Haven CT: Yale University Press, 1930.

Jakle, John A. *City Lights: Illuminating the American Night.* Baltimore: Johns Hopkins University Press, 2001.

Jones, Christopher F. *Routes of Power: Energy and Modern America.* Cambridge MA: Harvard University Press, 2014.

Jonnes, Jill. *Empires of Light: Edison, Tesla, Westinghouse, and the Race to Electrify the World.* New York: Random House, 2003.

Joseph, Gilbert M., Catherine C. LeGrand, and Ricardo D. Salvatore, eds. *Close Encounters of Empire: Writing the Cultural History of U.S.–Latin American Relations.* Durham NC: Duke University Press, 1998.

Joseph, Gilbert M., and Daniela Spenser, eds. *In from the Cold: Latin America's New Encounter with the Cold War.* Durham NC: Duke University Press, 2008.

"Juan Carosio, A Big Modern Hydro-Electric Power Station in Peru." *Brown Boveri Review* 26, no. 10 (October 1939), 227–34.

"Juan Carosio–Moyopampa Hydroelectric Station in Peru," *Brown Boveri Review* 42, no. 10, (October 1955): 405–13.
Kanduza, Ackson M. "'Let There Be Light': The Struggle for Developing Electricity Supply in Botswana, 1950–1970." *Botswana Notes and Records* 41 (2009): 39–46.
Kent, Robert B. "Geographical Dimensions of the Shining Path Insurgency in Peru." *Geographical Review*, 83, no. 4 (October 1993): 441–54.
Klarén, Peter. *Peru: State and Nationhood in the Andes*. New York: Oxford University Press, 2000.
Kofas, Jon V. *Foreign Debt and Underdevelopment: U.S.-Peru Economic Relations, 1930–1970*. Lanham MD: University Press of America, 1996.
Kuczynski Godard, Pedro Pablo. *Memoria de Energía y Minas 1980–1982*. Tomo, 1. Lima: Ministerio de Energía y Minas, 1982.
———. *Peruvian Democracy under Economic Stress: An Account of the Belaúnde Administration, 1963–1968*. Princeton NJ: Princeton University Press, 1977.
Lagendijk, Vincent. *Electrifying Europe: The Power of Europe in the Construction of Electricity Networks*. Amsterdam: Amsterdam University Press, 2008.
Landry, Marc. "Water as 'White Coal.'" RCC *Perspectives*, no. 2 (2012): 7–12.
La obra de los ingenieros en el progreso del Perú, Vol. 4. Lima: Editorial Perú Moderno, 1934.
Lara, Eduardo G. *Historia del Ministerio de Fomento y Obras Públicas, 1896–1936*. Lima: Imprenta y Librería del Gabinete Militar, 1935.
Larkin, Brian R. "The Politics and Poetics of Infrastructure." *Annual Review of Anthropology* 42 (2013): 327–47.
Larsen, Harold. *Preliminary Report on Peru*. Washington DC: World Bank, 1948.
Larson, Brooke. *Trials of Nation Making: Liberalism, Race, and Ethnicity in the Andes, 1810–1910*. Cambridge UK: Cambridge University Press, 2004.
Lasarte, Ezequiel. *Obras de represamiento en las lagunas de Huarochirí*. Lima: Impr. La Equitativa, 1920.
Latham, Michael E. *Modernization as Ideology: American Social Science and "Nation Building" in the Kennedy Era*. Chapel Hill: University of North Carolina Press, 2000.
———. *The Right Kind of Revolution: Modernization, Development, and U.S. Foreign Policy from the Cold War to the Present*. Ithaca NY: Cornell University Press, 2011.
Leonard, Thomas M., and John F. Bratzel, eds. *Latin America during World War II*. New York: Rowman & Littlefield, 2006.
Lilienthal, David E. *The Journals of David E. Lilienthal*. Vol. 6, *Creativity and Conflict,1964–1967*. London: Harper and Row, 1976.
Little, Arthur D. *A Program for the Industrial and Regional Development of Peru: A Report to the Government of Peru, 1960*. Cambridge MA: Arthur D. Little, Inc., 1961.

Lizarme Villcas, Nashely. "Education and Progress: The Escuela de Ingenieros and Engineers' Training in Peru (1911–1930)." *History of Education* (March 2025): 1–17.

Lizarme Villcas, Nashely, and Elías Amaya Núñez. "Energías del Centenario: ingenieros, industria y el sector energético en el Perú (1900–1920)." In *Desarrollo y sociedad en el Perú del Centenario*, edited by Patricia Palma and Jorge Lossio, 17–29. Lima: Pontificia Universidad Católica del Perú–Instituto Riva Agüero, 2022.

Llosa Larrabure, Jaime. "Cooperación Popular: un nuevo enfoque del desarrollo comunal en el Perú." *Revista Internacional del Trabajo* (September 1966): 289–90.

Long, Norman, and Bryan Roberts. *Miners, Peasants and Entrepreneurs: Regional Development in the Central Highlands of Peru*. Cambridge UK: Cambridge University Press, 1984.

———, eds. *Peasant Cooperation and Capitalist Expansion in Central Peru*. Austin: University of Texas Press, 1979.

López-Ocón, Leoncio. "La Sociedad Geográfica de Lima y la formación de una ciencia nacional en el Perú Republicano." *Terra Brasilis* 3 (2001), https://journals.openedition.org/terrabrasilis/330.

López Soria, José Ignacio, ed. *El pensamiento fascista en el Perú, 1930–1945*. Lima: Mosca Azul–Francisco Campodónico F., 1981.

———. *La Sociedad de Ingenieros del Perú: primera década, 1898–1908*. Lima: Universidad Nacional de Ingeniería, 2003.

Lowenthal, Abraham, and Cynthia McClintock, eds. *The Peruvian Experiment Reconsidered*. Princeton UK: Princeton University Press, 1976.

Ludeña, Sixto Elias. "Prospecting in the Peruvian Andes," *Ohio State Engineer* 12, no. 3 (1929): 8–9, 26.

Lust, Jan. *La lucha revolucionaria: Perú, 1958–1967*. Barcelona: RBA, 2013.

Mallon, Florencia E. *Peasant and Nation: The Making of Postcolonial Mexico and Peru*. Berkeley: University of California Press, 1995.

Mann, Michael. "The Autonomous Power of the State: Its Origins, Mechanisms and Results." In *States in History*, edited by John A. Hall, 109–36. Oxford UK: Blackwell, 1986.

Manrique, Nelson. *Mercado interno y región: la sierra central, 1820–1930*. Lima: DESCO, 1987.

Mariátegui, José Carlos. *Ideología y política*. Lima: Biblioteca Amauta, 1975.

———. *Siete ensayos de interpretación de la realidad peruana*. Lima: Biblioteca Amauta, 1968.

Martínez de la Torre, Ricardo. *Apuntes para una interpretación marxista de historia social del Perú*. Lima: Impresora Editora Peruana, 1947.

Martínez de la Torre, Ricardo, "La teoría del crecimiento de la miseria aplicada a nuestra realidad," *Amauta*, no. 27, (November–December 1929): 73–80.

Martuccelli Casanova, Elio. "Lima, capital de la Patria Nueva: el doble Centenario de la Independencia en el Perú." *Apuntes* 19, no. 2 (2006): 256–73.

Masterson, Daniel. *Fuerza Armada y sociedad en el Perú moderno: un estudio sobre relaciones civiles militares, 1930–2000*. Lima: Instituto de Estudios Políticos y Estratégicos, 2001.

Matos Mar, José. *Migration and Urbanization: The "Barriadas" of Lima: An Example of Integration into Urban Life*. Santiago: UNESCO, 1958.

McCook, Stuart. *States of Nature: Science, Agriculture, and Environment in the Spanish Caribbean, 1760–1940*. Austin: University of Texas Press, 2002.

Melgar Lazo, Hector. *Trazos para la historia de la cooperativa eléctrica*. Edited by Luis Carlos Arroyo. Huancayo: PICEISAC Ediciones, 2002.

Méndez, Cecilia. "Incas sí, indios no: apuntes para el estudio del nacionalismo criollo en el Perú." *Documento de trabajo 56*. Lima: Instituto de Estudios Peruanos, 2000.

Ministerio de Energía y Minas. *Energía eléctrica y desarrollo: plan de acción*. Lima: Ministerio de Energía y Minas, 1982.

———. *Exposición del Ministro de Energía y Minas*. Lima: Oficina de Relaciones Públicas, 1970.

———. *Mantaro*. Lima: Ministerio de Energía y Minas, 1973.

———. *Minería y energía en el Perú: política y situación actual*. Lima: Ministerio de Energía y Minas, 1977.

———. *Política de electricidad de la revolución peruana*. Lima: Ministerio de Energía y Minas, División de Publicaciones, 1972.

———. *Siete años de revolución en el sector energía y minas*. Lima: Oficina Sectorial de Planificación del Ministerio de Energía y Minas, 1976.

Ministerio de Fomento y Obras Públicas. *Ley de industria eléctrica No. 12378, reglamento y anexo sobre sistema de clasificación uniforme de cuentas*. Lima: Imprenta Atlántida, 1956.

Ministerio de Fomento y Obras Públicas, Dirección de Aguas e Irrigación. *Padrón de fuerza motriz hidráulica*. Lima: Imprenta Torres Aguirre, 1926.

———. *Prescripciones reglamentarias para la redacción y presentación de proyectos de irrigación, represamiento, aprovechamiento hidro-motriz, encauzamiento, defensa y demás obras hidráulicas*. Lima: Imprenta Americana, 1938.

Miró Quesada, Óscar. *Elementos de geografía científica del Perú*. Lima: Imprenta El Comercio, 1919.

Mitchell, Timothy. *Rule of Experts, Egypt, Techno-Politics, Modernity*. Berkeley: University of California Press, 2002.

Monsalve Zanatti, Martin. "Evolution of the Peruvian Large Family Business 1896–2012." In *Evolution of Family Business: Continuity and Change in Latin America and Spain*, edited by Paloma Fernández Pérez and Andrea Lluch, 238–54. Cheltenham UK: Edward Elgar Publishing, 2016.

Montaño, Diana. *Electrifying Mexico: Technology and the Transformation of a Modern City*. Austin: University of Texas Press, 2021.

Morton, David L. "Reviewing the History of Electric Power and Electrification." *Endeavour* 26, no. 2 (June 2002): 60–63.

Mumford, Jeremy Revi. *Vertical Empire: The General Resettlement of Indians in the Colonial Andes*. Durham NC: Duke University Press, 2012.

Muñoz Cabrejo, Fanni. *Diversiones públicas en Lima 1890–1920: la experiencia de la modernidad*. Lima: Red para el Desarrollo de las Ciencias Sociales en el Perú, 2001.

Murillo Garaycochea, Percy. *Historia del APRA: 1919–1945*. Lima: Editorial Atlántida, 1976.

Murra, John V. *El mundo andino: población, medio ambiente y economía*. Lima: Instituto de Estudios Peruanos–Pontificia Universidad Católica del Perú, 2002.

Naqvi, Ijlal. *Access to Power: Electricity and the Infrastructural State in Pakistan*. New York: Oxford Academic, 2022.

Natella, A. A. "Enrique Solari Swayne and Collacocha." *Latin American Theatre Review* 4, no. 2 (Spring 1971): 39–44.

National Rural Electric Cooperative Association. *Helping Others Help Themselves*. [n.p.]: NRECA, 1969.

Navarro Talavera, Emilio. *La electricidad en el Perú*. Lima: [s.n.], 1985.

Needham, Andrew. *Power Lines: Phoenix and the Making of the Modern Southwest*. Princeton NJ: Princeton University Press, 2014.

North, Lisa. "Orientaciones ideológicas de los dirigentes militares peruanos." In *El gobierno militar: una experiencia peruana 1968–1980*, edited by Cynthia McClintock and Abraham F. Lowenthal, 271–300. Lima: Instituto de Estudios Peruanos, 1985.

Oficina Nacional de Desarrollo de los Pueblos Jóvenes. *Anuario 1969–1970*. Lima: ONDEPJOV, 1970.

Orlove, Benjamin. "Putting Race in Its Place: Order in Colonial and Postcolonial Peruvian Geography." *Social Research* 60, no. 2 (Summer 1993): 301–36.

Ortiz, Fernando. *Cuban Counterpoint: Tobacco and Sugar*. Durham NC: Duke University Press, 1995.

Palacios Rodríguez, Raúl. *La Sociedad Geográfica de Lima: fundación y años iniciales*. Lima: Universidad de Lima, 1988.

Palavicino Fernández, Amanda. "Labor de la educadora familiar en el campamento Campo Armiño de la Central Hidroeléctrica del Mantaro." Undergraduate dissertation, Pontificia Universidad Católica del Peru, 1973.

Palomo, J. I. M. "Embracing the Cordillera: Luis Carranza Ayarza and the Development of Environmental Imaginaries in Late-Nineteenth-Century Peru." *História, Ciências, Saúde-manguinhos* 30 (2023): 1–20.

Paredes Hernández, Carlos Luis. "Arquitectura y discurso simbólico en el Oncenio de Leguía: una aproximación a través de sus edificios oficiales (1919–1924)." *Diacrónica* 4, no. 3 (2015): 33–47.

Pareja Paz Soldán, José. *Curso de geografía del Perú: lecciones dictadas en la Universidad Católica del Perú*. Lima: Librería e imprenta Gil, 1937.

———. *Geografía del Perú*. Vol. 1. Lima: Librería Internacional del Perú, 1950.

———. *Geografía del Perú*. Vol. 2. Lima: Librería Internacional del Perú, 1950.

———. *Geografía del Perú: curso universitario*. Lima: Libr. e Imp. D. Miranda, 1943.

———. *Geografía del Perú: manual*. Lima: Lib. Internacional del Perú, 1955.

Parker, David S. *The Idea of the Middle Class: White-Collar Workers and Peruvian Society, 1900–1950*. University Park: Pennsylvania State University Press, 1998.

Parker, Jason C. *Hearts, Minds, Voices: US Cold War Public Diplomacy and the Formation of the Third World*. Oxford UK: Oxford University Press, 2016.

Pastor Campos, Gonzalo C. "Peru: Monetary and Exchange Rate Policies, 1930–1980." IMF Working Papers 2012/166. International Monetary Fund, 2012.

Paz Soldán, Mariano Felipe. *Diccionario Geográfico Estadístico del Perú*. Lima: Imprenta del Estado, 1877.

Pease García, Henry. *El ocaso del poder oligárquico: lucha política en la escena oficial 1968–1975*. Lima: DESCO, 1977.

———. "La reforma agraria peruana en la crisis del estado oligárquico." In *Estado y política agraria: 4 ensayos*, edited by DESCO, 13–136. Lima: DESCO, 1977.

———. *Los caminos del poder: tres años de crisis en la escena política*. Lima: DESCO, 1979.

———. *Un perfil del proceso político peruano: a un año del segundo belaundismo*. Lima: DESCO, 1981.

Pennano, Guido. "Desarrollo regional y ferrocarriles en el Perú." *Apuntes* 5, no. 9 (1979): 131–51.

Pepper, Charles M. "Electricity in Peru," *Electrical Review* 52, no. 8 (February 1908): 319–20.

Perú. Dirección Nacional de Estadística y Censos. *I Volumen de resultados de los censos nacionales: sexto censo nacional de población*. Lima: Dirección Nacional de Estadística y Censos, 1965.

Perú. Oficina Nacional de Estadística y Censos. *Censos nacionales, VII de población, II de vivienda, 1972*, vol. 15: *Departamento de Lima*. Lima: Oficina Nacional de Estadística y Censos, 1974.

Peru President. *A Program of Action: Address delivered by General Manuel A. Odría, Constitutional President of Peru, before a joint session of Congress, on the inauguration of his administration, July 28, 1950*. Lima: Dirección General de Informaciones del Perú: 1950.

Perú Presidente. *El Perú construye: mensaje presentado al Congreso Nacional por el Presidente Constitucional de la República, Arquitecto Fernando Belaúnde Terry, el 28 de julio de 1968*. Lima: Minerva, 1968.

Perz, Stephen George, and Jorge Luis Castillo Hurtado. *The Road to the Land of the Mother of God: A History of the Interoceanic Highway in Peru*. Lincoln: University of Nebraska Press, 2023.

Pike, Frederick. *The Modern History of Peru*. New York: Praeger, 1969.

"Plan de electrificación—Disposiciones Oficiales, Resolución Suprema No. 13, Lima 18 June 1951," *Electrotécnica*, no. 3 (January 1953): 18–20.

Planas Silva, Pedro. *El 900: aproximaciones al 900*. Lima: CITDEC, 1994.

———. *La descentralización en el Perú republicano, 1821–1998*. Lima: Municipalidad Metropolitana de Lima, 1998.

Plasencia Soto, Rommel. "El cambio socioeconómico en las comunidades de la sierra central: una revisión." *Investigaciones Sociales*, año 4, no. 6 (2000): 127–40.

Poole, Deborah. *Vision, Race, and Modernity: A Visual Economy of the Andean World*. Princeton NJ: Princeton University Press, 1997.

Portocarrero, Felipe. *El imperio Prado, 1890–1970*. Lima: Universidad del Pacífico, 1995.

Portocarrero S., Felipe, and Luis Camacho S. "Impulsos moralizadores: el caso del Tribunal de Sanción Nacional 1930–1931." In *El pacto infame: estudios sobre la corrupción en el Perú*, edited by Felipe Portocarrero, 35–73. Lima: Red para el Desarrollo de las Ciencias Sociales en el Perú, 2005.

Portocarrero, Gonzalo. *De Bustamante a Odría: el fracaso del Frente Democrático Nacional, 1945–1950*. Lima: Mosca Azul, 1983.

Portocarrero C., Juan N. *Contribución al estudio de los recursos hidráulicos para fuerza motriz en el Perú*. Lima: Imp. Torres Aguirre, 1920.

Posnansky, Arthur. *A los pueblos de Argentina, Bolivia, Chile y Perú. La independencia económica de Sud América: una central eléctrica internacional, aprovechando el desnivel del Desaguadero hacia el Pacífico*. La Paz: Editorial Renacimiento, 1937.

Pratt, Mary Louise. *Imperial Eyes: Travel Writing and Transculturation*. New York: Routledge, 1992.

Pretel, David, Ian Inkster, and Helge Wendt. "Technology in Latin American History: Perspectives, Scales and Comparisons." *History of Technology* 34 (2019): 1–22.

Puente, Javier. "Making Peru's Sendero Luminoso: The Mega Niño of 1982–1983." *Age of Revolutions*. https://ageofrevolutions.com/2017/03/27/making-perus-sendero-luminoso-the-mega-nino-of-1982-3/.

———. *The Rural State: Making Comunidades, Campesinos, and Conflict in Peru's Central Sierra*. Austin: University of Texas Press, 2023.

Pulgar Vidal, Javier. *Geografía del Perú*. Lima: PEISA, 1996.
Purcell, Fernando. "Imaginarios socioculturales de la hidroelectricidad en Sudamérica 1945–1970." *Atenea* 518 (2018): 97–116.
"*¿Que es Cooperación Popular?*" Lima: [n.p.], 19.
Quiñones Tinoco, Leticia. *Construir y modernizar: el Ministerio de Fomento, 1896–1930*. Lima: Universidad Nacional de Ingeniería, 2014.
Quiroz, Alfonso. *Corrupt Circles: A History of Unbound Graft in Peru*. Baltimore: Johns Hopkins University Press, 2008.
Ráez y Gómez, Nemesio Augusto. "El Mantaro y sus afluentes." *Boletín de la Sociedad Geográfica de Lima*, vol. 12 (1902): 161–84.
Ramírez Alzamora Cobos, Claudio. *Santiago Antúnez de Mayolo: vida y Obra*. Lima: ELECTROPERU, 1980.
Ramírez Villacorta, Yolanda. "La penetración capitalista en una comunidad campesina: el caso de San Pedro de Casta, Huarochiri." *Debates En Sociología* 5 (1980): 39–70.
Rankin, William J. "Infrastructure and the International Governance of Economic Development, 1950–1965." In *Internationalization of Infrastructures: Proceedings of the 12th Annual Conference on the Economics of Infrastructures*, edited by Jean-François Auger, Jan Jaap Bouma, and Rolf Künneke, 61–75. Delft: Delft University of Technology, 2009.
República Peruana. *Diario de los debates de la Cámara de Diputados: legislatura ordinaria de 1957*, vol. 4. Lima: Congreso del Perú, 1958.
———. *Diario de debates de la Cámara de Diputados: legislatura ordinaria de 1959*, vol. 1. Lima: Congreso del Perú, 1960.
———. *Diario de los debates de la Cámara de Diputados: legislatura ordinaria de 1960*, vol. 2. Lima: Congreso del Perú, 1961.
———. *Diario de los debates de la Cámara de Diputados: segunda legislatura extraordinaria de 1960*, vol. 3. Lima: Congreso del Perú, 1961.
———. *Diario de los debates de la Cámara de Diputados: legislatura ordinaria de 1961*, vol. 3. Lima: Congreso del Perú, 1962.
———. *Diario de los debates de la Cámara de Diputados: legislatura ordinaria de 1961*, vol. 4. Lima: Congreso del Perú, 1962.
———. *Diario de los debates de la Cámara de Diputados, legislatura ordinaria de 1964*, vol. 6. Lima: Congreso del Perú, 1965.
Rice, Mark. *Making Machu Picchu: The Politics of Tourism in Twentieth-Century Peru*. Chapel Hill: University of North Carolina Press, 2018.
———. "Roads to Progress: Public Press: Public Perceptions of Highway Conceptions of Highway Construction in Peru, 1920–30." New York: CUNY Academic Works, 2015.
Roberts, Richard. *Schroders: Merchants & Bankers*. London: Palgrave Macmillan, 1992.

Rodríguez Valencia, Katya, and Lizardo Seiner Lizárraga. *Juan Alberto Grieve Becerra y Ricardo Tizón y Bueno*. Lima: Universidad Nacional de Ingenieria, 2000.

Romero Padilla, Emilio. *El descentralismo*. Lima: Tarea, 1987.

———. *Geografía económica del Perú*. Lima: Imprenta Torres Aguirre, 1939.

———. *Historia económica del Perú*. Lima: Fondo Editorial de la UNMSM, 2006.

Rose, Mark H. *Cities of Light and Heat: Domesticating Gas and Electricity in Urban America*. University Park: Pennsylvania State University Press, 1995.

Rosell, Ricardo García. "La irrigación de la costa del Perú." *Boletín de la Sociedad Geográfica de Lima*, Vol. 3 (1893): 121–46.

Rosenstein-Rodan, P. N. "Problems of Industrialisation of Eastern and South-Eastern Europe." *Economic Journal* 53, no. 210/211 (1943): 202–11.

Rostow, W. W. *The Stages of Economic Growth: A Non-Communist Manifesto*. Cambridge UK: Cambridge University Press, 1960.

Rowe, Leo Stanton. *Early Effects of the War upon the Finance, Commerce and Industry of Peru*. New York: Oxford University Press, 1920.

Rubio, Daniela. "Las guerrillas peruanas de 1965: entre los movimientos campesinos y la teoría foquista." *Histórica* 32, no. 2 (2008): 123–67.

Salvatore, Ricardo D. *Disciplinary Conquest: U.S. Scholars in South America, 1900–1945*. Durham NC: Duke University, 2016.

Schivelbusch, Wolfgang. *Disenchanted Night: The Industrialization of Light in the Nineteenth Century*. Berkeley: University of California Press, 1995.

Schmidt, Gregory D. "Political Variables and Governmental Decentralization in Peru, 1949–1988." *Journal of Interamerican Studies and World Affairs* 31, no. 1/2, Special Issue: Latin America at the Crossroads: Major Public Policy Issues (Spring–Summer 1989): 193–232.

SCIPA. *Desarrollo agrícola y económico de la zona del Mantaro en el Perú*. Lima: Ministerio de Agricultura–SCIPA, 1954.

Scott, James C. *Seeing Like a State: How Certain Schemes to Improve the Human Condition Have Failed*. New Haven NC: Yale University Press, 1998.

Scott Palmer, David. *The Shining Path of Peru*. New York: Palgrave Macmillan 1994.

Sharp, Daniel. *U.S. Foreign Policy and Peru*. Austin: University of Texas Press, 1972.

Sheahan, John. *Searching for a Better Society: The Peruvian Economy from 1950*. University Park: Pennsylvania State University Press, 1999.

Sheffield, Glenn Francis. "Peru and the Peace Corps, 1962–1968." PhD Dissertation, University of Connecticut, 1991.

Smith, Neil. *Uneven Development: Nature, Capital, and the Production of Space*. New York: Blackwell, 1984.

Smith, Philip Seabury. "Electrical Goods in Peru and Ecuador." In *United States Department of Commerce, Special Agent Series, No. 154*, 1–51. Washington DC: Government Printing Office, 1917.

Sneddon, Christopher. *Concrete Revolution: Large Dams, Cold War Geopolitics, and the US Bureau of Reclamation*. Chicago: Chicago University Press, 2015.

Sobrevilla Perea, Natalia. "¿Qué tan falaz fue la prosperidad?: auge y caída del estado guanero (1840–1880)." In *La condena de la libertad: de Túpac Amaru II al bicentenario peruano en seis ensayos y un colofón*, edited by Paulo Drinot and Alberto Vergara, 95–142. Lima: Crítica: Universidad del Pacífico, Fondo Editorial, 2022.

Sociedad Geográfica de Lima. *Cincuentenario de la Sociedad Geográfica de Lima*. Lima: Librería e Imprenta Gil, 1937.

Soifer, Hiliel. *State Building in Latin America*. Cambridge UK: Cambridge University Press, 2015.

Solari Swayne, Enrique. *Collacocha*. Lima: UNMSM, 1992.

Stiglich, Germán. *Geografía comentada del Perú*. Lima: Casa Editora Sanmartí & Cía., 1913.

St. John, Ronald Bruce. *The Foreign Policy of Peru*. Boulder: Lynne Rienner Publishers, 1992.

Strange, P. "Two Electrical Periodicals: The Electrician and The Electrical Review 1880–1890." *IEE Proceedings. Part A, Physical Science, Measurement and Instrumentation, Management and Education* 132, no. 8 (1985): 574–81.

Stubbs, Ricardo Walter. *Registro histórico de Lurigancho y Chosica*. Lima: Edit. Médica Peruana, 1958.

Suazo, Miguel, "La Central hidroeléctrica Santiago Antúnez de Mayolo." *El Ingeniero Civil* 1, (May–June 1979): 14–22.

Sulmont, Denis. *El movimiento obrero en el Perú: 1900–1956*. Lima: PUCP. Departamento de Ciencias Sociales, 1975.

Swyngedouw, Erik. *Liquid Power: Contested Hydro-Modernities in Twentieth Century Spain*. Cambridge MA: MIT Press, 2015.

Tarazona-Sevillano, Gabriela. "The Organization of Shining Path." In *The Shining Path of Peru*, edited by Scott Palmer, 189–208. New York: Palgrave Macmillan, 1994.

Téllez Contreras, León Felipe. "Infrastructural Politics: A Conceptual Mapping and Critical Review." *Urban Studies* 62, no. 1 (2024): 31–51.

Thorp, Rosemary, and Geoffrey Bertram. *Perú: 1890–1977. Crecimiento y políticas en una economía abierta*. Lima: Universidad del Pacífico, 2013.

———. *Peru, 1890–1977: Growth and Policy in an Open Economy*. London: Macmillan Press, 1978.

Tischer, Julia. *Light and Power for a Multiracial Nation. The Kariba Dam Scheme in the Central African Federation*. Basingstoke: Palgrave Macmillan, 2013.

Tizón y Bueno, Ricardo. *La reorganización del Ministerio de Fomento*. Lima: Impr. del Centro Editorial, 1914.

Turner, W. T. *Las lagunas de Huarochirí y su futuro ensanche*. Lima: Litografía y Tip. Carlos Fabbri, 1908.

Ueda Tsuboyama, Augusto Martín. *Historia del Cuerpo de Ingenieros de Minas del Perú, 1902–1950*. Lima: Universidad Nacional de Ingeniería, 2002.

Ulloa y Sotomayor, Alberto. *Posición internacional del Perú*. Lima: Atlántida, 1977.

UNICEF. *Servicios básicos integrados en áreas urbano-marginales del Perú*. Lima: UNICEF, 1981.

United Nations Framework Convention on Climate Change. "How Hydropower Can Help Climate Action." November 21, 2018. https://unfccc.int/news/how-hydropower-can-help-climate-action.

United States Bureau of Foreign and Domestic Commerce. *Electrical Development and Guide to Marketing of Electrical Equipment in Peru*. Washington DC: U.S. Government Printing Office, 1927.

United States Congress. *Foreign Assistance Act of 1964, Hearings before the Committee on Foreign Affairs, House of Representatives. Part 1*. Washington: U.S. Government Printing Office, 1964.

———. *Implementation of the Humphrey Amendment to the Foreign Assistance Act of 1961: Third Annual Report to the Congress*. Washington DC: U.S. Government Printing Office, 1964.

Universidad Nacional de Ingeniería. *Foro sobre el proyecto hidroeléctrico del río Mantaro, del 27 al 31 de octubre de 1964*. Lima: Universidad Nacional de Ingeniería, 1965.

Valdez Arroyo, Flor de María. *Las relaciones entre el Perú e Italia (1821–2002)*. Lima: PUCP, 2004.

Vallenas, Fritz. *La ley peruana de electricidad*. Lima: Asociación de Empresarios Eléctricos del Perú, 1959.

———. "La Ley de la Industria Eléctrica y sus consecuencias." *Electrotécnica*, no. 17–18 (July–December 1956): 13–18.

Velíz Lizárraga, Jesús. *El Perú y la cultura occidental*. Lima: Instituto de Investigaciones Sociales del Perú, 1957.

vom Hau, Matthias. "State Infrastructural Power and Nationalism: Comparative Lessons from Mexico and Argentina." *Studies in Comparative International Development* 43, nos. 3–4, (2008): 334–54.

vom Hau, Matthias, and Valeria Biffi. "Mann in the Andes: State Infrastructural Power and Nationalism in Peru." In *Peru in Theory*, edited by Paulo Drinot, 191–216. Oxford UK: Palgrave Macmillan, 2014.

Walter, Richard J. *Peru and the United States, 1960–1975: How Their Ambassadors Managed Foreign Relations in a Turbulent Era*. University Park: Pennsylvania State University Press, 2010.

Werlich, David. *Peru: A Short History*. Carbondale: Southern Illinois University Press, 1978.

Whitaker, John. *Americas to the South*. New York: The Macmillan Company, 1939.

Willimetz, Emil, dir. *Light of the Andes (Luz de los Andes)*. 1969. Produced by Audio Visual Productions. S.A. Sponsored by the United States Information Service. Film reel. 15 min.

Winner, Langdon. "Do Artifacts Have Politics?" In *Technology and Society: Building Our Sociotechnical Future*, edited by Deborah G. Johnson and Jameson M. Wetmore, 209–226. Cambridge MA: MIT Press, 2009.

Wittfogel, Karl. *Oriental Despotism: A Comparative Study of Total Power*. New Haven CT: Yale University Press, 1957.

Wolfenson, Azi. *El gran desafío*. Lima: Intergráf. de Servicios, 1981.

Yauri Montero, Marcos. "Deidades panandinas del Perú antiguo en el Callejón de Huaylas (Ancash): 'Los Dioses De Pumakayan.'" SCIENTIA 18, no. 18 (2016): 59–70.

Zitor. *Historia de las principales huelgas y paros obreros habidos en el Perú, 1896–1946*. Lima: [n.p.], 1976.

Zolov, Eric. "Introduction: Latin America in the Global Sixties." *The Americas* 70, no. 3 (2014): 349–62.

INDEX

Acción Popular (AP), 129, 140, 148, 163, 175–76, 183, 203–4
Adams, George, 72–73
Alabama Power Company, 84
Alayo River, 142, 244n53
Alfaro de la Peña, Víctor, 148
Alianza Popular Revolucionaria Americana (APRA), 88, 100, 132–33, 140, 143, 156, 176, 180, 183, 217; attacks on infrastructure by, 65, 207–8; and attitudes towards fascism, 64, 77–79; and unions, 64–66
Alliance for Progress, 11, 15, 127, 129, 139, 140–41, 143, 153, 174, 190
Amazon: economic potential, 6, 20, 22, 24, 27, 83, 165, 172
Amazon River, 24, 28, 116
anarchism, 59
Ancash, 89, 141, 166, 169
Antúnez de Mayolo, Santiago, 10, 39–40, 42, 80, 99, 102, 164, 171, 187, 195, 200–201, 203; and involvement in politics, 11, 169–70, 181–82; surveys by, 166–69
Apata, 151
Apurímac River, 27
Arana, Víctor, 33, 42, 62
Arequipa, 35, 46, 103, 209
Aristocratic Republic, 27, 60, 63, 188, 214; and bureaucratic development, 36–37; economic conditions during, 19, 56; and international migration, 26, 30, 50; labor disputes during, 59–60; technical education during, 32–34
Arkansas Electric Cooperatives, Inc., 150, 155, 159
Arthur D. Little Inc., 172–73
Association of Electrical Businessmen (AEE), 94, 96, 181
Asunción Tramways Light and Power, 61
Ayacucho, 33, 165, 173, 179

Balta, José, 72
Bambarén, Luis, 198
Banco Italiano, 56, 60, 78, 82–83
Banco Popular del Perú, 56
Barbablanca, 73
Barbillion, Louis, 167
Basadre, Jorge, 19, 51
Baudin, Louis, 130–31
Beals, Carlton, 78
Belaúnde Terry, Fernando, 1, 128, 140, 143, 147–48, 153–54, 159, 160–61, 163–64, 183–85, 187–88, 195; intellectual views of, 121–22, 127, 129–31, 135, 203–4, 213
Belmont Bar, Alejandro, 106–7
Beltrán, Pedro, 174–76, 179, 193
Benavides, Óscar, 66, 68, 76–79
Benso, Camilo (Count of Cavour), 22
Bianchini, Gino, 61, 63, 65–67, 69, 76–79, 81, 83–85
Billinghurst, Guillermo, 60
Bisso Loredo, Carlos, 177
Black, Eugene, 101, 111–12
Blanco Montesinos, Alicia, 107
Boggio, Bartolomeo, 55
Bolsheviks, 45, 65
Boner, Pablo, 10, 76, 81, 83–85, 99, 102, 111, 118, 120, 193, 195, 203, 214; surveys by, 54, 69–74, 116
Boveri, Walter E., 61, 77, 83–84, 111–13, 120

Brown, Boveri & Company, 61, 76, 84, 112, 114, 120
Bryce, Murray D., 172
Buch, Alfred, 171
Buse, Hermann, 73–74, 115–17
Bustamante y Rivero, José Luis, 84, 91, 99–100, 237n12

Caballero Alvarado, Luis, 184
Callahuanca Power Plant, 69, 79, 84–85, 117, 195, 210; construction of, 73–75; inauguration of, 76–77; survey and design of, 73
Callao, 38, 42, 55, 62–63, 67, 186
Calmet, Mario, 127, 144–45, 148
Cañón del Pato Hydroelectric Plant. *See* Santa Corporation
Carosio, Juan, 60–61, 67, 69–70, 76–78, 84, 184
Carranza, Luis, 20
Castillo, Pedro, 211
Castro Pozo, Hildebrando, 130, 136
Centro de Altos Estudios Militares (CAEM), 190
Cerro de Pasco (CPC), 35, 170–71, 173, 177, 183, 188–89, 199
Chanduví Torres, Luis, 65
Chauca, 73
Chaufournier, Roger, 113–14
Chimbote, 80, 103
China, 176–77, 211
Chiozza Money, Leo, 34–35
Chosica, 53, 57–59, 65, 71
Chosica Power Plant, 57–60, 62, 64, 73, 79, 88
Christian Democrats (DC), 107, 175–76, 178
Civilista Party, 19, 60, 98, 238n33
Cold War, 5, 7–9, 11, 91, 130, 153; guerrillas in Peru during, 148–49, 154, 184; and linkages to infrastructure, 8–10, 141–44, 149, 159–60, 163–64, 174, 179, 184, 215; role of experts during, 11, 127–28, 132, 140–45, 150, 160; and theories of development, 8–10, 12, 129, 160–61, 163–64, 174–77, 179, 190, 192, 215, 240n76. *See also* Alliance for Progress
Communist Party of Peru, 63. *See also* Shining Path
Compañía Eléctrica del Callao, 55
Compañía Ítalo-Argentina de Electricidad, 61
Compañía Nacional de Tranvías, 68
Compañía Sudamericana de Electricidad (SUDELECTRA), 61, 64, 112
Concepción, 137, 153, 205
Concepción Power Plant, 138, 140, 144, 154
Cooperación Popular, 131, 139, 143, 153
Cooperativa Eléctrica Comunal del Centro Ltda. No. 127, 15, 127, 140, 199; expansion of, 148–51, 154–56; formation of 142–44; and interaction with U.S. specialists, 140–45, 150–51, 155; and local practices, 128, 142, 145–46, 155, 161; internal political dynamics of, 156, 158–60; and relationship with the state, 144–45, 149–50, 154–56, 158–61; role of Peace Corps in, 144–48, 159; social dynamics of, 145–47, 155; social impact of, 151, 153. *See also* cooperativism
cooperativism: history in Peru of, 131–32, 136; intellectual discourses on, 129–35. *See also* Cooperativa Eléctrica Comunal del Centro Ltda. No. 127
Coppo, Dionigi, 195
Cornejo Chávez, Héctor, 107, 175
Coronel Zegarra, Enrique, 24
Corporación de Energía Eléctrica del Mantaro (CORMAN), 178, 180–82, 185–89, 196, 200, 209. *See also* Mantaro Hydroelectric Plant
Cuerpo de Minas, 37
Cunas River, 142

Cuzco, 33, 35, 103, 105, 141, 168, 173

Decentralist Party, 47
decentralization, 5, 14, 18, 44–52, 96, 109, 138, 164–65, 173–76, 179–81, 199–201, 212
de Habich, Eduardo, 33, 37
de la Puente Uceda, Luis, 148, 184
de la Riva-Agüero y Osma, José, 26–27, 48, 81
de Lavalle, Hernando, 83, 93
Delgado, Eulogio, 24
Dependency Theory, 99, 106, 164, 174, 190, 193
development: elite views on, 9, 89–90, 92, 98–100, 103–5, 107–8, 110–11, 123–24, 164–65, 173–79, 187, 216–17; and linkages to infrastructure, 6, 8–9, 98–103, 105–8, 124, 163–65, 171–77, 190, 192, 215–16; and planning, 8, 90, 97–98, 101–5, 170–74, 192, 203, 249n48; theories of, 8–10, 90, 109, 122–24, 129, 132, 160–61, 163–65, 174–77, 179, 190, 215
Dibós Dammert, Eduardo, 113
Dimond, Francis, 140–41, 143–44, 148–49
Dnieper Hydroelectric Station, 45
d'Ornellas, Tómas, 33
Durand, Augusto, 53, 59, 66

Edison, Thomas, 10
electric appliances, 39, 42–43, 62
Électricité de France (EDF), 102, 138, 171
Electroconsult, 181, 185
ELECTROLIMA, 199, 206
ELECTROPERU, 192–93, 203–4, 207, 209
elites: as economic groups; 9, 89–93, 97–98, 100–101, 105–6, 108, 110, 124, 174–75, 179; regional, 13, 51–52, 138, 149, 164, 173–74, 181, 183–86, 201; and social attitudes towards electricity, 40–43; and views on development of, 9, 89–90, 92, 98–100, 103–5, 107–8, 110–11, 123–24, 164–65, 173–79, 187, 216–17; and views on indigenous populations, 2, 5, 17–20; 24–28, 48–51, 117–20, 122, 124, 213–14; and views on the Andes, 5–7, 17, 20–28, 48–51, 122, 124, 129–31, 209, 212–14; and views on urbanization, 5, 89–90, 93–94, 98, 108–9, 123–24, 164–65, 173, 180, 198–200
Ellis, Clyde, 141, 159
Empresa Comercializadora de Energía Eléctrica de Muquiyauyo (FEBO), 137, 149
Empresas Eléctricas Asociadas (EE. EE. AA.). *See* Lima Light and Power Company (LL&P)
Empresa Transmisora de Fuerza Eléctrica, 55
Energía Hidroeléctrica Andina (HIDRANDINA), 84, 145
engineers: and involvement in politics, 3, 11, 63, 66–68, 77–79, 81, 96–97, 169–70, 181–82, 214; and local knowledge, 54, 71–72, 102, 164, 166, 200, 215, 239n52; promotion of hydroelectricity by, 30, 32–33, 59, 99, 102, 181–82; scientific excellence from the periphery by, 10, 54, 102, 164, 166, 200, 216; surveying of hydraulic sources by, 34, 37, 69–74, 102–3, 115–16, 138, 142, 167–70; and travels through Peru, 17–18, 28–29, 32, 58, 70–72, 168; and views on technical education, 32–33
English Electric Company, 178
Enock, Charles Reginald, 28–32, 210
Espantoso, Guillermo, 55

fascism: influence on electric companies, 11, 14, 34, 54, 61, 66–67, 77–79, 80, 85, 88, 112, 215; in Peru, 64, 77–79, 81, 83

Fernández Maldonado, Jorge, 154–55, 189–90, 195, 197
First World War, economic impact in Peru of, 60
Flores Galindo, Alberto, 7
Flusin, George, 167
Fondo Nacional de Desarrollo Económico (FNDE), 138
France: capital flows from, 138, 140, 155, 170; engineers from, 90, 102–3, 138, 171
Franco, Francisco, 83
Freundt Rosell, Víctor, 107, 187
Fujimori, Alberto, 209–10
futurism, 81

Gallagher, Manuel, 83, 93
Gálvez, José, 40, 230n88
García Calderón, Francisco, 25–28, 49, 51, 120, 199, 213–14
García Rosell, Ricardo, 22
Garland, Alejandro, 23
Gazzani, Fernando, 66–68
Generación del 900, 25–26, 40, 44, 49
General Confederation of Peruvian Workers, 64
geography of Peru: and foreign engineers, 28–32, 69–70, 72; and hydraulic potential, 5–7, 14, 17–18, 20, 22–24, 28–30, 47–51, 69–72, 98–99, 116, 212–13; intellectual views on the, 5–6, 14, 17–20, 22–27, 44–50, 98–99, 115–18, 121–22, 197, 212–13; and race, 19–20, 25–28, 49, 117–20, 122, 124, 213–14
George Wimpey and Co. Limited, 178
Germany, 61, 82, 134, 164, 167, 203; capital flows from, 178, 181; engineers from, 171; relations with Peru, 178, 184–85; technological flows, 84
GIE-Impregilo, 184–85, 187, 190
Gottschalk, Alfred L. M., 29–30
Great Britain, 34–35; capital flows from, 56–57, 78, 83, 149, 164, 178, 181; engineers from, 28–32; relations with Peru, 178, 184–85; technological flows from, 36, 58, 139
Great Depression, 4, 8, 14, 54, 62–63, 66–67, 84, 144
Grieve, Juan Alberto, 63, 100
Grieve Madge, Jorge, 100–101, 103, 108, 176–77, 193, 239n42
Guarini, Emilio, 30–33, 35, 37, 59, 166–67
Guevara, Ernesto "Che," 148
Guevara, Víctor, 107
Gutiérrez, Sixto, 148

Haya de la Torre, Víctor Raúl, 65, 79, 132
Henry Schroder & Co. Investment Bank, 56–57, 64
Hirschman, Albert, 6
Huamalí, 137–38, 140
Huampaní Power Plant, 69, 85, 88
Huancavelica, 1, 12, 38, 107, 140, 179, 185–87, 205, 207, 209, 211; hydraulic potential in, 165, 168–70
Huancayo, 35, 136–37, 139–40, 144, 147–48, 155–56, 159, 163, 168, 171, 185–86; elites from, 13, 142, 149, 164, 173–74, 181–84; intellectuals from, 132–35
Huancayo Industrial Society, 136–37
Huarisca Power Plant, 138, 140, 144, 155
Huarochirí Lake System 12, 71–74
Huatica River, 55
Huaura River, 182
Huinco Power Plant, 14–15, 69, 85–86, 89–91, 105, 109, 124–25, 195, 203, 207; construction of, 116–20; financing of, 111–15; inauguration of, 121–23; survey and design of, 115–16
Humphrey, Hubert, 141, 148
Hunting Associates Limited, 172

Hydroelectricity: and attacks on infrastructure, 1, 53, 65, 88, 205–8, 211, 217; and bureaucratic development, 8, 35–38, 51, 99–100, 106–8,

124, 144–45, 154, 188–89, 214; and decentralization debates, 5, 14, 18, 44–52, 96, 109, 138, 164–65, 173–76, 179–81, 199–201, 212; and geography, 6–7, 17–18, 22–24, 26–30, 47–51, 69–72, 82, 98–99, 115–16, 197, 212–13; ideological ramifications of, 14, 45, 54, 65, 79–81, 88, 176, 197, 212; and industrial development, 8–9, 22–23, 45, 89–90, 95–99, 101–3, 105–9, 163–64, 172–77, 179–80, 192, 216; and interconnectivity, 2, 15, 140, 158, 160–61, 171, 192, 199, 201, 205, 208–10; and legislation; 35–36, 90, 92–97, 100, 105, 113, 158, 192–93, 204, 209; and linkages to agrarian reform, 5, 156, 164, 173, 175, 179, 216; and local knowledge, 11–12, 54, 71–72, 102, 145–46, 161, 164, 166, 200, 215, 239n52; and nationalism, 79, 122, 155–56, 175, 188–89, 195, 197–98; and race, 7, 30–31, 117–20, 122, 124, 213–14; social impact of, 18, 39–42, 73–74, 76, 116–17, 151, 153, 186, 198; and state planning, 8, 90, 97–98, 101–5, 170–74, 192, 203, 249n48; and theories of development, 8–10, 90, 109, 122–24, 160–61, 163–65, 174–77, 179, 190, 192, 215; and urbanization, 39–40, 89–90, 93–95, 98, 108–9, 123–24, 164–65, 173, 176, 180, 197–200, 238n19

Import Substitution Industrialization (ISI), 106, 154, 188, 192, 200
Inca Empire, 12, 110, 142, 163, 165–66, 168–69, 170, 200, 210, 213–14; intellectual interpretations of, 7–8, 23–25, 29, 46–48, 129–31, 134–35, 197; and linkages to hydroelectric works, 121–22, 146
Indigenismo, 44–46, 48–49, 120
indigenous populations: impact of infrastructural projects on, 73–74, 76, 116–17, 151, 153, 186; and internal migration, 5, 89–90, 93–94, 108–9, 122–24, 164–65, 173, 180, 197–200; lima elites views on, 2, 5, 17–20, 24–28, 48–51, 117–20, 122, 124, 213–14; and local knowledge, 11–12, 54, 71–72, 145–46, 161, 215; regional intellectuals views on, 44–46, 51, 132–35, 160; role in infrastructural projects of, 73–76, 90–91, 117–20, 122, 124, 131, 139, 145–46, 185–86, 197, 213–14; and technical education, 18, 26, 32–33, 51
industrialization, and linkages to electricity, 8–9, 22–23, 45, 89–90, 95–99, 101–3, 105–9, 163–64, 172–77, 179–80, 192, 216
Ingenio Power Plant, 136–38, 140, 154
Instituto Nacional de Cooperativas (INCOOP), 131, 143, 148
Instituto Nacional de Planificación (INP), 102, 180
International Bank for Reconstruction and Development (IBRD), 8, 15, 90, 96, 101, 105, 109–15, 120, 124–25, 178, 193, 203
International Petroleum Company (IPC), 79, 91, 148–49, 153, 155, 175, 183, 189, 197
Isola, Gio Batta, 56
Italy: capital flows from, 11, 14, 60–61, 76, 78, 81–83, 164, 184–85, 197, 211; engineers from, 30–33, 37, 54, 59–61, 63, 65–69, 76–79, 83–85, 166–67, 181; and relations with Peru, 77–79, 81–83, 195, 197
Izquierda Unida (United Left), 208

Japan, 82, 105, 118, 120, 171
Jasper-Newton Electrical Cooperative, 141
Jauja, 137, 140, 168, 173, 185, 227n18
Jochamowitz, Alberto, 73

Johnson, Lyndon B., 149
Jones, Barton, 80
Jones, Wesley, 148
Junín, 12–13, 89, 107, 165, 179; communist activities in, 148–49, 154, 184; hydraulic potential in, 12, 71, 111–12, 116, 136–40, 142–43, 183; regional elites in, 13, 127, 138, 149, 173–74, 181, 183–86, 201. *See also* Huancayo; Mantaro Valley
Junta Nacional de Vivienda, 198

Kemmerer Mission, 66–67
Kennedy, John F., 129
Kenyon, A. L., 58
Klein Mission, 92, 111
Korean War, 95
Kuczynski Godard, Pedro Pablo, 204

Labarthe, Abel, 102–3
labor: and Marxist influence in unions, 14, 54, 63–66; and strikes, 59–60, 63–69
Lake Junín, 163, 183
Lake Titicaca, 31–32, 163
Landauer, Carl, 101
Lara, Eduardo, 38
Latina Lux Company, 61
Law of Industrial Promotion, 106
Leguía, Augusto, 33–35, 39, 53, 60–63, 66–68, 100. See also *Oncenio*
Lenin, Vladimir, 45, 79, 176, 197
Liberal Party, 53
Lilienthal, David, 249n48
Lima: blackouts in, 1, 53, 59–60, 88, 206–8; centennial celebrations in, 40, 61; economic and political dominance of, 44–45, 49, 51–52, 90, 109, 115, 123–24, 164–65, 179–80, 199–200, 212; elite views on, 5, 40–41, 89, 93–94, 98, 108–9, 123–24, 164–65, 173, 179–80, 199–200; hydraulic potential in, 12, 45, 57–58, 69, 70–73, 85, 92, 111, 193, 195; and social impact of electricity, 40–43; urban growth of, 39, 44, 89–90, 93–95, 108–9, 123–24, 164–65, 173, 176, 180, 197–200
Lima Light and Power Company (LL&P), 26, 53–54, 89–90, 108–9, 145, 173, 177, 203, 206, 209, 211; and company life, 85; and impact on local communities, 73–74, 76, 116–17, 155, 241–42n99; and international organizations, 14–15, 84, 90, 109, 111–15, 120, 124–25, 193, 195; and labor relations, 54, 59–60, 63–69, 85; and linkages to fascism, 11, 14, 54, 61, 66–67, 77–81, 83–85, 88, 112, 215; origins of, 55–57; and relationship with the state, 60–68, 92–93, 95, 99–100, 113–15, 182–84, 187–89, 193, 195, 198–99; tramway division of, 55–56, 59, 62, 64–65, 68–69; transnational economic linkages of, 56–57, 61, 64, 66–68, 76–78, 81–84, 112, 114, 195. *See also* Callahuanca Power Plant; Huinco Power Plant
Lircay, 38–39
Ludeña, Elias, 71–72

Machu Picchu Power Plant, 105, 168, 185
Mann, Michael, 4, 106
Mantaro Hydroelectric Plant, 1, 12, 15, 105, 108–9, 138, 144, 158, 200–201, 209–10; attacks on, 205–7; construction of, 181, 185–90, 195; financing of, 177–79, 181, 184–85; inauguration of, 195–97; political debates surrounding, 163–65, 174–79, 181–88; surveying and design of, 168–73. *See also* Corporación de Energía Eléctrica del Mantaro (CORMAN)
Mantaro River, 12, 27, 116, 136–37, 142, 163, 165–66, 168–70, 201, 203

Mantaro Valley, 12–13, 15, 205–6, 217; electrification of, 127, 136–40, 142–44, 148–51, , 153–56, 160; hydraulic potential in, 142–44; intellectual views on, 132–35. *See also* Cooperativa Eléctrica Comunal del Centro Ltda. No. 127
Manzanares, 142
Marcapomacocha Lake System, 12, 71, 89, 111–12, 115–18
Marcona Mining Company, 91, 173, 177
Mariátegui, José Carlos, 44–46, 48–49, 51, 63–64, 66, 130, 136, 199, 208, 213
Mariotti, Carlos, 84, 99, 112–13, 117, 120, 155, 182, 188, 193, 195, 198–99
Martinelli Tizón, Augusto, 99, 113, 139
Martínez, Manuel Alonso, 139
Martínez de la Torre, Ricardo, 64, 66–67
Marxism: and association with hydroelectricity, 45, 54, 176, 197; and attacks on infrastructure, 65, 88, 205–8, 217; and labor unions, 14, 54, 63–66; Peruvian interpretations of, 45, 48–49, 79, 130, 136; and theories of development, 8, 9, 127, 153–54, 163–64, 174
Mary, Marcel, 102
Maryknoll Society, 131
Matucana Power Plant, 69, 193, 195
McClellan, Daniel, 131
Melgar Casallo, Ponciano, 134–35
Mendoza Gálvez, Alfonso, 185
Meriwether Lewis Electric Cooperative, 141
Military junta of 1962, 102, 121, 129, 180
mining legislation, 35, 91, 97
Ministry of Energy and Mines, 8, 38, 154, 188, 198
Ministry of Fomento and Public Works, 62, 72–74, 93–94, 96, 102, 127, 138–39, 148–50, 159, 167–69, 171–72, 181, 214; bureaucratic organization of, 8, 36–38, 51, 99–100, 106–8, 132, 136, 144–45, 154, 188–89; and international organizations, 113–14; and labor, 64, 66, 68
Miró Quesada, Alfonso, 179
Miró Quesada, Carlos, 78
Miró Quesada, Óscar, 17, 22–23
Mitchell, Troy, 141, 144
Mobile Oil, 184
Modernization Theory, 8–9, 12, 90, 109, 122–24, 129, 132, 163, 174–76, 179, 190, 215
Monge, Carlos, 49, 118
Montero Muelle, Alfonso, 181
Morales Bermúdez, Francisco, 158, 181, 200
Motor Columbus, 61, 84, 111–12
Movimiento Democrático Pradista (MDP), 97, 107–8, 176–77
Moyopampa Power Plant, 69, 84–85, 206
Muquiyauyo, 136–37, 149
Mussolini, Benito, 77–78

National Agrarian Society (SNA), 92, 95
National Electrification Plan, 14, 90, 97–98, 101–6, 108, 111, 113, 115, 123–24, 137–38, 171
National Institute for Cooperatives (INCOOP), 131, 143, 148
National Rural Electric Cooperative Association (NRECA), 127–28, 141–45, 148–51, 155, 159
National School of Arts and Crafts, 32–33, 36, 59
National Society for Industry (SNI), 34, 93, 105, 181
National Society of Mining and Oil, 181
National University of Engineering, 33, 100, 145, 178
Neoliberalism, 131, 161, 187, 193, 201, 203–4, 209, 211, 217
New Deal, 15, 54, 101, 128, 141, 150

Niagara Falls, 31, 80
Noriega Calmet, Fernando, 99, 100
Nozaki, Tsuguo, 171

Ochenio, 101; economic conditions during, 91–93, 95–97, 110–11, 138; urbanization during, 94
Odría Amoretti, Manuel, 89, 91–97, 102, 107, 111, 138, 169–70, 183, 237n12, 238n24. See also *Ochenio*
Office for the Development of Pueblos Jóvenes (ONDEPJOV), 198–99
Oficina Nacional de Desarrollo Cooperativo (ONDECOOP), 156
Oil: legislation, 91–92, 97, 204; and relationship to hydroelectricity, 24, 36, 47, 70, 76, 79, 122, 175, 180, 184, 189, 197. *See also* International Petroleum Company (IPC)
Oncenio, 39, 43, 51, 60–62, 78, 100; economic conditions during, 61; labor relations during, 63–64, 68; modernization of Lima during, 39–40, 61
Ortiz, Fernando, 133
Ouachita Electric Plant, 150

Pampas, 165, 185–86, 188, 190, 195, 206
Panedile, 185
Pardo, Manuel, 227n18
Pardo y Barreda, José, 59
Pareja Paz Soldán, José, 48–50
Pativilca Power Plant, 103, 105
Payán, José, 56
Paz Soldán, Mariano Felipe, 20, 226n10
Peace Corps (PC), 144–48
Peña Costa, Juan Manuel, 56
Pepper, Charles M., 32, 58
Peru and London Bank, 56
Peruvian Electrotechnical Association (AEP), 70, 94, 99–103, 181, 189, 205
Peruvian Technical Association, 81
Piaggio, Faustino, 55
Piazza y Valdez Ingenieros S. A., 120, 154
Portocarrero, Juan, 37
Prado, Mariano Ignacio, 97–98
Prado Ugarteche, Manuel, 86, 94, 121, 132, 167–68, 170, 174, 176, 187–88, 193, 195, 209; foreign policy of, 79–83, 88, 177–79, 190; and views on development, 90–91, 97–99, 100–103, 108, 111, 114, 124, 137–38, 163–64, 172, 179–80; and views on Lima, 98, 108–9, 164, 179–80, 198–99, 201, 212
Prado Ugarteche, Mariano, 55, 60, 72, 98
Prebisch, Raúl, 106, 193
Prialé, Ramiro, 183–84
Public Works Committees (JOP), 138
Pucará Power Plant (Junín), 139
Puente Revilla, Pedro, 189
Puno, 31, 33, 46, 131, 135

Ráez y Gómez, Nemesio Augusto, 165, 168, 195
Raimondi, Antonio, 20, 28, 47, 107, 165, 204
Recavarren, Jorge Luis, 179
Reddy Kilowatt, 84, 145
Revolutionary Government of the Armed Forces (GRFA), 5, 15, 38, 107, 117, 128, 160–61, 203; energy policy of, 158, 192–93, 195–99; foreign policy of, 154, 159, 188, 190, 200, 215; political divisions within, 158–59, 189, 200; urbanization policy of, 197–99; views on development, 153–54, 156, 158, 164, 188–90, 192
Revolutionary Left Movement (MIR), 148–49, 184
Rey de Castro López de Romaña, Jaime, 178
Rimac River, 12, 55, 58, 69–70, 72–73, 85, 115–16, 170, 241n96
Rizo Patrón, Alfonso, 113–14, 172, 178
Rochdale Pioneers, 130
Roldán, Alejandro, 62–63

Romero Padilla, Emilio, 23, 46–48, 50, 96, 231n106, 231n113
Roosevelt, Franklin D., 82
Rosenstein-Rodan, Paul, 109
Rostow, Walt, 8, 109, 122, 149, 174–76, 183, 240n76
Rural Electric Association (REA), 141, 150

Salocchi, Gino, 60–61, 77–78, 81, 83
Samamé, Mario, 178
Samanez Ocampo, David, 63
Sánchez Cerro, Luis Miguel, 62–63, 65–66
San Marcos University, 32–33, 38, 46, 78, 166–67
San Pedro de Casta, 73
Santa Catalina Industrial Society, 55
Santa Catalina Textile Factory, 55
Santa Corporation, 12, 80, 86, 89, 97, 103, 167–70 177–78, 207
Santa Eulalia River, 12, 58, 69–70, 72–74, 85, 111–12, 115–16
Santa River. *See* Santa Corporation
Santa Rosa Electric Company, 55, 57
Santa Rosa Power Plant, 55, 62, 64–65, 76, 199
Sarmiento Espejo, Juan, 107
School of Engineers, 32–33, 36, 81. *See also* National University of Engineering
scientific journals, 31–32
Scotland, 34–35
Second World War and Peru: and blacklisting, 82–83; and relations with the Axis powers, 14, 77–79, 81–83; and relations with the United States, 14, 54, 79–83, 88
Seoane, Manuel, 79
Serrano Solís, Mario, 140
Servicio Cooperativo Interamericano de Producción de Alimentos (SCIPA), 142, 151
Servicios Eléctricos Nacionales (SEN), 108, 155
Sheque Power Plant, 195, 203
Sheque Reservoir, 118, 120
Shining Path, 1–2, 205–9, 217
Sicaya, 145–47, 155
Siemens-Schuckertwerke, 178
Silva y Silva, María Eleonora, 107, 137
Sistema Eléctrico Interconectado Nacional (SEIN), 209
Sistema Nacional de Apoyo a la Movilización Social (SINAMOS), 156
Sobrevilla González, Luis, 139
Socialist Party of Peru, 63, 130
Sociedad de Alumbrado Eléctrico y Fuerza Motriz de Piedra Liza, 55
Sociedad Geográfica de Lima (SGL), 20, 22–25, 27, 37, 46–47, 73, 231n106
Société Swiss de Electricité Americaine, 112
Society of Engineers, 34, 71
SOCIMPEX, 138, 140, 155
Solari Swayne, Enrique, 89, 118
Solf y Muro, Alfredo, 79
Soviet Union, 45, 54, 65, 82, 101, 130, 176–77; relations with Peru, 154, 159, 188, 190, 200–201, 215
Stanton Rowe, Leo, 60
State Commission for Electrification of Russia (GOELRO), 45
Stiglich, Germán, 22
Stiles, Albert, 34, 73
Sutton, Charles, 34
Swiss Federal Institute of Technology, 60, 69
Switzerland: capital flows from, 11, 14, 54, 60–61, 76–78, 82–84, 112, 195, 211; engineers from, 10, 54, 69–74, 83–85, 111–13, 116–18, 120, 182, 188, 193, 198–99, 214; technological flows from, 76, 84, 111, 114, 120, 195, 211

Tahuantinsuyo. *See* Inca Empire
Tapia, Mario, 135
Tayacaja, 165, 168, 184–87, 195, 198–99, 206

Taylor, John, 144
technical education, 18, 26, 32–34, 51, 215
Tennessee Valley Authority (TVA), 54, 80–81, 88, 173
Tesla, Nikola, 10
Texas Electric Cooperatives, 150, 155
Thompson Whitaker, John, 77
Tidwell, Paul, 141
Tittman, Harold, 110
Tizón y Bueno, Ricardo, 34, 36
Trujillo, 35, 65, 103
Turner, William, 34, 73

Ugarteche, Pedro, 55
Unión Nacional Odriista (UNO), 143, 156
United Nations Economic Commission for Latin America (CEPAL), 96, 106
United States: capital flows from, 56, 61, 143–44, 148–49, 170; circuits of expertise from, 127–28, 132, 140–45, 159, 161, 215; and electric goods trading, 39, 42–43; engineers from, 32, 34, 54, 58–59, 72–73, 80; relations with Peru, 54, 78–83, 128–29, 148–50, 153–55, 159–60, 178–79, 190, 200, 211, 246n93; technological flows from, 58, 81, 84, 150–51, 154–55
United States Agency for International Development (USAID), 128, 131, 140–41, 143–44, 148–50
urbanization, 39, 44, 89–90, 93–95, 98, 108–9, 123–24, 164–65, 173, 176, 180, 197–200
U.S. Foreign Assistance Act of 1961, 141

Vaccari, Pietro, 61, 63, 68–69, 76–79, 83, 85
Vallenas, Fritz, 96–97, 99, 139
Vaughn, Jack, 148
Velasco Alvarado, Juan, 107, 158, 164, 189, 197, 199, 200, 213
Véliz Lizárraga, Jesús, 133–34
Vicos, 142
Vilcanota River, 168
Villareal, Federico, 166
von Humboldt, Alexander, 19, 28
von Mises, Ludwig, 131
Voorduin, William, 172

Walton Electric Cooperative, 144
War of the Pacific, 19, 25, 45, 56, 72–73, 98, 100, 155, 166
water legislation, 35
Watson Cisneros, Eduardo, 175
Weber, Max, 4
Werner-Gren, Axel, 80
Wiredhand, Willie, 145
Wittfogel, Karl, 7
Wolfenson, Azi, 203
World Bank. *See* International Bank for Reconstruction and Development (IBRD)
Wormeringer, Raymond, 102
Wunenburger, Gaston, 84, 99, 112, 120

Yanacoto Power Plant, 57–60, 62, 64–65, 69, 79, 84, 88
Yauli River, 35